T0113719

COLLABORATIVE INNOVATION

Bart Barthelemy

Wright Brothers Institute

BALBOA.PRESS
A DIVISION OF HAY HOUSE

Copyright © 2020 Bart Barthelemy.

All rights reserved. No part of this book may be used or reproduced by any means, graphic, electronic, or mechanical, including photocopying, recording, taping or by any information storage retrieval system without the written permission of the author except in the case of brief quotations embodied in critical articles and reviews.

This book is a work of non-fiction. Unless otherwise noted, the author and the publisher make no explicit guarantees as to the accuracy of the information contained in this book and in some cases, names of people and places have been altered to protect their privacy.

Balboa Press books may be ordered through booksellers or by contacting:

Balboa Press
A Division of Hay House
1663 Liberty Drive
Bloomington, IN 47403
www.balboapress.com
1 (877) 407-4847

Because of the dynamic nature of the Internet, any web addresses or links contained in this book may have changed since publication and may no longer be valid. The views expressed in this work are solely those of the author and do not necessarily reflect the views of the publisher, and the publisher hereby disclaims any responsibility for them.

The author of this book does not dispense medical advice or prescribe the use of any technique as a form of treatment for physical, emotional, or medical problems without the advice of a physician, either directly or indirectly. The intent of the author is only to offer information of a general nature to help you in your quest for emotional and spiritual well-being. In the event you use any of the information in this book for yourself, which is your constitutional right, the author and the publisher assume no responsibility for your actions.

Any people depicted in stock imagery provided by Getty Images are models, and such images are being used for illustrative purposes only. Certain stock imagery © Getty Images.

Print information available on the last page.

ISBN: 978-1-9822-4811-6 (sc)
ISBN: 978-1-9822-4813-0 (hc)
ISBN: 978-1-9822-4812-3 (e)

Balboa Press rev. date: 05/21/2020

CONTENTS

INTRODUCTION

Collaborative Innovation can produce powerful results and move teams and organizations from breakdowns to breakthroughs. Every team wants collaboration and every organization wants innovation. Yet there are thousands of books, seminars and articles produced each year that focus on how to help you achieve these capabilities. So combining the two should be even harder to achieve. Surprisingly, not so! What we found is that Innovation needs Collaboration and that Collaboration needs Innovation. Although we started to explore both separately, we discovered that they are synergistic. Actually, both are necessary if you really want to achieve transformative results and new breakthroughs.

This book captures some of the steps and paths that the Wright Brothers Institute (WBI) took on its journey to Collaborative Innovation. While it certainly focuses on the best practices that we found, the book also describes some of the lessons that we learned and miscues that we made. The good news is that the reader doesn't have to repeat everything that we did to achieve Collaborative Innovation. There are practices and principles that will be discussed that can be used to stimulate, facilitate, support and sustain Collaborative Innovation. But don't be fooled, it will take effort and commitment. Collaborative Innovation is both and art and a science and it's mostly about how people behave when they really collaborate and innovate.

Collaborative Innovation didn't start with the Wright Brothers Institute. It has been around for millennia, but the creativity and efforts of the people who have worked at WBI over the past twenty years and their supporters, partners and customers have taken Collaborative Innovation to a new level. This is their story and there are so many to thank that it's not possible to name them all. However, any institute must be created, renewed and nurtured for it to succeed, so great thanks to the Founding Colleagues: Vince Russo, Bill Bogert, Dan Curtiss and Bob May for their vision and guidance; the succeeding Directors: Les McFawn, Dave Walker and Wendell Banks for their wonderful leadership and hard work; our terrific supporters: Mick Hitchcock, Alok Das, Dave Shahady, Jeff Graley, Joe Sciabica, Morley Stone, Scott Galster, Ricky Peters, Doug Ebersole, Bill Harrison and Jack Blackhurst; and all of the men and women who have been part of the WBI team. While leaders are important, nothing happens without doers and WBI could not have succeeded without Rich Maresca, Diane Sowar, Martin Steiger, Bob Mulcahy, Craig Steffen, Diane Williams, Tina Speers, Heidi Susta, Jennie Hempstead, Anne Hertenstein, Angie Pickard, Mike Osgood, Jim Masonbrink, Rob Klees, Chris Remillard, Bob Lee, Emily Riley, Cheryl Reed, Steve Fennessey and Scott Springer. Some of the WBI pioneers in collaboration and innovation are mentioned specifically in the following sections, but anyone who has been a part of the Wright Brothers Institute has made it and the world better.

▎ THE BEGINNING

"It's always best to start at the beginning", Glinda, Wizard of Oz

On a snowy Saturday morning in January 2002, five guys met in a small conference room near Wright Patterson Air Force Base in Dayton, Ohio. It would be nice to say that this was a diverse group, but it wasn't. Just five middle-age men who had spent most of their lives in or near the Air Force Research Laboratories and Development Centers on the Base. They were senior executives who felt that their mission in life was to do the research and develop the technology that would make the United States Air Force not only superior to any other Air Force in the world, but several generations of technology ahead of the nearest competitor. All had spent most of their service careers focusing on the potential catastrophe that would occur if the United States and the Soviet Union went from a cold to a hot war. Fortunately, that had been averted and, while probably not the only reason, the United States' technological superiority certainly was a factor in preventing World War 3.

After September 11, 2001, it was clear that the need to pursue and develop superior Air Force technology was not over. The new enemies might be different but they still had to be deterred or stopped if the United

States was to be secure. And the new twenty-first-century technologies, the internet, smart phones, artificial intelligence and social networks, were already emerging and had to be incorporated into the development of Air Force technology. But there was a problem. What had worked in the twentieth century would not suffice for the new age. A bureaucratic, structured and stove-piped Air Force Research and Development system would not produce the capabilities that would be needed to stay ahead of our competitors. In addition, global companies like Google, Microsoft, and Amazon would be spending far more than the Air Force on R&D. Constant innovation would be required to stay ahead of countries, organizations and individuals who would go to any length, including suicide, to destroy our way of life. A new way of doing business for the Air Force was required.

But changing a bureaucratic and highly regulated structure, controlled by Washington, and steeped in the cultural norms of the past would take a lot of time and effort and still might not happen. So the five guys invented something different, an organization that would help the Air Force be collaborative and innovative by being a partner, actually an intermediary, to the Air Force's R&D Centers at Wright-Patterson Air Force Base.

Within one year we were up and running, we had secured the name Wright Brothers Institute from the Wright family in Dayton, and had persuaded the Air Force to give us both the funds and a contract to start operating. From the beginning, we invented, developed, explored and adapted ways to help the Air Force's Research and Development organizations at the Base with collaboration and innovation.

This is the story of that journey and the emergence of the powerful process that we call Collaborative Innovation.

| WRIGHT BROTHERS INSTITUTE

"Let's solve that flying problem", Orville and Wilbur Wright

Officially, the Wright Brothers Institute (WBI) is a non-profit organization (501c3) that has a very unique relationship with the Air Force Research

Lab (AFRL) called a Partnership Intermediary Agreement. This particular type of agreement was a last minute choice by the Air Force to get WBI going, transfer some funds to catalyze some collaborative and innovative activities and provide a simple way for the government to do business with an outside partnering and intermediary organization. Generally a government agency like the Air Force contracts with an outside organization for services and products and there is a very formal relationship between that agency and the outside company, university or other government entity. While the Partnership Intermediary Agreement is a contractual vehicle, it's more like a grant to an intermediary for support services to help the government organization achieve what they want. In this case, the intermediary was WBI and the support services that the AFRL wanted were collaborative environments, facilitation support to enhance innovation and partnering with non-traditional organizations.

| COLLABORATIVE ENVIRONMENTS

While cooperation and contractual agreements between the Air Force and any outside person, organization, university or company occur all the time, true collaboration can be difficult because of the myriad of regulations and acquisition rules that the Air Force must follow. Often a collaboration between AFRL and an outside organization is curtailed simply because there may be a perception of favoritism or pre-selection. That severely limits the Air Force's ability to converse, coordinate, cooperate and eventually collaborate with organizations that may eventually become contractors through a competitive acquisition process. In order to assist the Air Force R&D community, and particularly AFRL, WBI has created a variety of collaboration environment, spaces and facilities to enhance the Air Force's ability to collaborate with outside organizations. Interestingly, collaboration within the Air Force R&D community can also be difficult because of geographical separation, bureaucratic structures, mission segregation and cultural norms that do not reward multi-disciplinary collaboration.

In order to maximize the effectiveness of WBI's collaborative environment support, significant research was conducted on collaboration approaches that had become best practices for similar organizations. For example, the Santa Fe Institute provides supportive environments and stimulating approaches to encourage Nobel Laureates to engage and communicate on very challenging global problems. The spaces in the Santa Fe Institute are divided by large whiteboards and glass walls and their guests are encouraged to post ideas on them with non-permanent markers to encourage collaboration and innovation. There are mandatory morning and afternoon all-hands sessions to insure at least two hours of conversation each day, enhanced by coffee and donuts in the morning and wine and cheese in the afternoon.

Many visits were also taken to existing government and corporate collaboration environments to understand their best practices, like rapidly reconfigurable furniture and settings, non-intrusive power and communication connections and a wide variety of unique social-media, conferencing, recording and display software and hardware systems to enhance collaboration. All of these were incorporated into the WBI environments and they were enhanced yearly when new ideas or technologies became available.

While collaboration environments are only one piece of the Collaborative Innovation puzzle, they are critical to its achievement. In the examples given later in this book, their role in catalyzing and facilitating collaboration, innovation and Collaborative Innovation cannot be underestimated.

INNOVATION SUPPORT

The innovation process in the Air Force is extremely complex. The Air Force's competition comes from any real or potential enemy of the United States. Not only do the threats change rapidly but much of what these competitors are doing is highly secretive, even more covert than intellectual property. The systems that the Air Force develops and supports are also incredibly complex. A fighter jet is made up of engines that have over 10,000 unique parts, aircraft skins that are subjected to extreme environments but also must be stealthy, ISR (Intelligence, Surveillance, Reconnaissance) and communication systems that are constantly upgraded with the latest technologies, and human performance systems that must support the pilots and crews in extreme environments. In addition, the Air Force R&D community must plan and execute a balanced portfolio that will enable innovative, transformational and force-multiplier technology development to satisfy immediate, near-term, mid-term and far-term needs.

WBI began by trying to find and understand the most effective innovation approaches being used by industry, academia and government organizations. As time went on, we not only used the latest in Front End and Back End Innovation techniques and practices but constantly modified and adapted them for use with the AFRL and Air Force R&D community. Soon we realized that to really help the Air Force, we had to develop, prototype and apply some unique and new approaches to innovation while constantly staying abreast of the expanding set of innovation practices and processes being used by other organizations. Over the past fifteen years, one of WBI's primary goals has been to provide the Air Force with innovation support, services and consultation to help them stay at the leading edge of innovative technology development. Many of the processes and tools described later in this book are examples of these efforts.

▌ PARTNERSHIP INTERMEDIARY

As the name implies, WBI serves as an intermediary between AFRL and other organizations that can assist AFRL in unique ways. The Air Force already has a myriad of processes and practices that connects them to the traditional industrial, business and academic organizations that assist the Air Force in carrying its mission. WBI supplements these approaches by acting as a catalyst, facilitator and energizer that helps AFRL connect with non-traditional partners, such as small-businesses, non-profits, new industrial and academic partners and even individuals who would find it difficult to communicate with a highly bureaucratic government organization. This WBI role is consistent with its mission to provide innovation and collaboration services to the Air Force since unique ideas and novel partnerships come from these connections. Over the past fifteen years, WBI has served as an intermediary between AFRL and thousands of organizations, businesses and individuals that have brought new possibilities, opportunities and approaches to AFRL.

COLLABORATION

"I can do things you cannot do, you can do things I cannot do, together we can do great things." Mother Theresa

▍COLLABORATION AND COMPETITION

There's an organizational model which suggests that most business relationships progress through several stages: they start with Communication, if appropriate, they move to Coordination, sometimes they result in Cooperation, occasionally they lead to Competition, but it's truly special if they end up in Collaboration. During WBI's first five years, the primary emphasis was on trying to achieve or improve collaboration, both inside and outside of AFRL.

When AFRL was formed in 1998, the organization was structured around ten major Technical Directorates focusing on technology areas that dominated the Air Force's research program. Five of the Technical Directorates were located at Wright-Patterson Air Force Base, along with the AFRL headquarters. The other five were based in Washington DC, New Mexico, New York and Florida. While there were occasional examples of cross-Directorate coordination and cooperation, there was little collaboration between the Technical Directorates. Further, there was little collaboration within the Technical Directorates since they were organized bureaucratically in Divisions and Branches. Occasionally, S&Es from a Branch might reach out to other S&Es within the Directorate, but rarely was there even coordination between S&Es in different Directorates. There was clearly overlap and even unintended duplication between the research and development programs in the different Directorates. Not surprisingly, the Technical Directorates behaved like they were in competition with one another, certainly when it came to funding, manpower and other resources. While the AFRL leadership preached and promoted collaboration for the first five years of AFRL's existence, there was very little collaboration within AFRL when WBI was formed. Based in Dayton near Wright-Patterson Air Force Base, we focused our attention on generating collaboration between and within the Technical Directorates at Wright-Patterson AFB.

On the outside, it was just as bad. Because of the government contracting rules and the need for real and perceived fairness in spending taxpayers' money, the industrial, business and academic organizations which provided substantial support to AFRL (over 75% of AFRL's $2-4 Billions/year funding went to outside contractors) were in high competition to win contracts and receive AFRL funding. Occasionally, AFRL would form a collaboration with a university to create a Center of Excellence in a technical area or a Collaborative Research and Development Agreement (CRADA) was developed, usually with a company, to jointly develop technology without the exchange of any funds. But those transactions were exceptions, the rule was to issue requirements through a Request for Proposal (RFP), evaluate the responses and select a winner for the project or program. Once selected, a contractor would guard their relationship with AFRL to the exclusion of even communications with other potential contractors or collaborators. This led to a tremendous amount of duplication of effort in the proposal phase of the process since everyone wanted to win and not surprisingly, each developed much of the same capabilities and expertise to win the effort. Not only did this eliminate any benefits of a potential collaboration, AFRL and their contractor base were entwined in very formal and legally restricted relationships which significantly reduced the potential of innovation or even the new ideas and approaches being considered. While WBI worked collaboration within AFRL, we also spent a significant amount of energy on stimulating and energizing potential collaborations between the industrial and academic partners and the contractors of AFRL.

GOVERNMENT AND INDUSTRY COLLABORATIONS

As WBI went about creating environments to support collaboration meetings and workshops, we decided to take on one of the toughest Air Force issues: how to get their three largest industry providers to collaborate. Back in the nineties, Norm Augustine, then CEO of Martin Marietta Corporation, convinced the then ten major Air Force prime contractors to merge into three companies. In his book, "Augustine's Laws", he argued that the potential business opportunities with the Air Force would either stabilize or decline in the future and it would not be sufficient to satisfy the business needs of all ten separate companies. After months of discussions and negotiations, three mega-companies emerged: Boeing, Lockheed-Martin and Northrop-Grumman. While Boeing did not have a hyphenated name, it actually was a merger of Boeing Aerospace and McDonnell-Douglas.

Interestingly, Norm Augustine consulted with us as he formed the Lockheed-Martin Corporation through a merger of several smaller companies. At one point, he asked for advice on how he could get the leaders of the now-merged smaller companies to not only co-operate with each other under the Lockheed-Martin banner but to get them to actually collaborate. While we had no magic formula, we did urge him to try using a Talking Stick to get the conversation and communications started. We had used a Talking Stick in a number of different situations to get individuals and teams to open-up in an honest and forthcoming way and it had actually worked. Our Talking Stick had been crafted by a Shaman jeweler and friend (Ross LewAllen) in Santa Fe, New Mexico and, while just a beautiful piece of carved wood about three feet long, it looked quite mystical. Ross would only carve a Talking Stick if the owner would agree to a session with him on a mountain in Santa Fe and would abide by two simple rules while using the Talking Stick, "to Hear from the Heart and to Speak from the Soul". In use, the Talking Stick is passed from one person to another, the holder must speak openly and truthfully and all others must listen very carefully to the words of the holder. Norm called Ross, went to Santa Fe, got his Talking Stick and used it as he formed the very complex and powerful Lockheed-Martin Corporation.

While the Talking Stick had helped Norm Augustine form one of the most powerful companies in America, collaboration between the three major conglomerates rarely occurred at that time. In actuality, it was competition for Air Force business that drove their behavior and each company would spend incredible amounts of money to win a big Air Force contract. Once the winner was chosen, the other companies might get a small piece of the contract but it was typically "winner take all". While competition was necessary and had many good features, the investments by each company during the proposal stage of most Air Force procurements were extremely large and obviously found their way into what the Air Force paid for its systems. In an attempt to see if collaboration between the three primes was possible, we convened a meeting at Wright Brothers Institute to discuss the subject. All three companies and the Air Force sent several of their senior leaders to the meeting and Wright Brothers Institute facilitated the discussion.

While many areas of potential collaboration were discussed, one topic which received significant support by all three companies and the Air Force was pre-competitive simulations of potential solutions to needed Air Force capabilities. Modeling and simulation of concepts and early prototypes is necessary before any proposals or decisions can be made to go forward. Actually constructing and testing physical prototypes that meet the needs identified by the Air Force is cost and time prohibitive and the Air Force will not fund the competitors to do that prior to selection and award. Nevertheless, proposals from the companies to justify their concept need to contain sufficient evidence that the system can be built and that it will perform as designed. For example, modeling and simulation of a new airplane design in a real or virtual wind tunnel satisfy this need. But the cost of developing sophisticated models of new designs and "flying" them in wind tunnels is expensive. Each of the major Air Force suppliers have extensive teams and facilities to do this and much of the pre-competitive effort in this area is redundant. If there were shared simulations and facilities that these companies could use, the proposal costs could be significantly reduced. The issue is to maintain the competition and guard each companies' intellectual property in such a shared venture. Because of the potential cost and time savings, each company was interested in pursuing this possibility.

After many meetings and considerable scientific, legal and financial discussions, agreement was reached to share some of the capabilities and facilities of the companies for both their common good and that of the Air Force. One of the catalysts that helped with the company collaboration was agreement by the Air Force to share its capabilities and facilities with all of companies in the pre-competitive phase. Since the Air Force's modeling and simulation capabilities are used to analyze and evaluate the competitor's submissions, this also allowed the companies to gain insight into the parameters and specifications that would be of most interest to the Air Force during the competitive phase. In essence, it was a win-win for all and a breakthrough that would result in significant savings for the companies, the Air Force and ultimately the nation. While the details of this collaborative process were ironed out over several years, collaboration was achieved in an area where it was not expected. All agreed that Wright Brothers Institute's role as a committed and persistent neutral facilitator was a key contributor to the success of this Collaborative Innovation.

GOVERNMENT AND UNIVERSITY COLLABORATIONS

Much of the basic and early research that provides the Air Force with the breakthrough technologies that create and enhance their systems and products comes from universities. Universities generally have research programs that are aligned with their departments and funding for these research efforts often comes from the government. Research professors and their teams are primarily rewarded for publications in peer-reviewed journals and single disciplinary symposiums and these publications are very important in achieving tenure and maintaining funding. While collaboration between researchers in the different university departments is increasing with the development of university institutes and multidisciplinary focus areas, it was not that common fifteen years ago. Very little multi-disciplinary research was being conducted in universities and very little cross-university or industry collaborative efforts aimed at specific Air Force related capabilities were occurring. In order to enhance the benefit and application of university research for the Air Force as well as to guide university research towards Air Force needs, WBI took on an initiative to enhance the collaboration of specific universities and selected industrial research laboratories with the Air Force Research Laboratory's Technical Directorates at Wright-Patterson Air Force Base.

We started with a technical area of strategic importance to the Air Force, nano-materials technology, and a nearby university, the University of Dayton. While the University of Dayton (UD) was already supporting AFRL in many technology development areas, we wanted to form a world-class Center of Excellence in nano-materials research at UD which would include collaborations with several nationally recognized nano-materials universities and research centers and be closely aligned with the Air Force's needs and programs in nano-materials technology. It took us about a year but we obtained three million dollars of matching funds to hire a world-class nano-materials researcher who would serve as an Endowed Chair at UD to head up the Center of Excellence. Commitments were also obtained from several universities, companies and AFRL to fund a significant multi-year program in this research area and a co-operative agreement between UD and the Air Force was initiated to provide research fellowships and education to the AFRL scientists. The Center of Excellence was a major achievement for WBI, AFRL and UD as well as a unique learning experience for the creation of government and university collaborations.

❙ GOVERNMENT, INDUSTRY AND UNIVERSITIES

Because the Air Force Research Laboratory is the Air Force's primary R&D organization, the Air Force relies on AFRL to accomplish its technology development mission. To satisfy this responsibility, AFRL must either lead, leverage or watch the development of any near, mid or far-term technology that might be of interest to the Air Force. In some cases, AFRL actually conducts much of the basic, exploratory and advanced development of critical tactical and strategic technologies with contractual support from industry and academia. In a few military critical areas, AFRL serves as the primary developer of these technologies, for example, hypersonic flight, directed energy weapons and advanced aircraft propulsion. In other areas, AFRL focuses on staying aware of what global and national industry and academia are developing and watches to see if any of these technologies need to be incorporated into Air Force current or future systems. A particularly interesting strategy is to leverage technologies that are being developed by industry, academia and other government laboratories for integration into Air Force capabilities. Not only does this give the Air Force a bigger bang for their buck but it also necessitates forming collaborations with these organizations in order to maximize the payoff of these technologies to the Air Force as well as the non-Air Force organization. Under the PIA, WBI can serve as an intermediary between AFRL and other organizations for the collaborative development of important technologies. From its inception, WBI has played this role in many critical technology areas and has often launched these collaborations though the use of Deep Dives, many of which have led to Collaborative Research and Development Research Agreements (CRADAs).

WBI has utilized Deep Dives as a form of Collaborative Innovation in order to explore, understand, examine and construct potential collaborations between various parties interested in developing mutually beneficial technologies. The first Deep Dive that WBI conducted for AFRL was in the area of advanced visualization technologies for military-critical decision making. While Deep Dives generally include a multi-day workshop where interested parties come together to discuss the chosen technology area, the process starts with a thorough and global examination of what organizations are involved in the development of the technologies of interest. In the case of advanced visualization technologies, an extensive search was conducted for several weeks and an initial global landscape of this technology area was developed. This allowed both the research and the researchers to be identified and provided the Deep Dive organizers with a list of potential participants to be invited to a workshop to explore collaboration. A three-day workshop with over fifty government, academic and industry participants was conducted with facilitation provide by WBI. With significant pre-work, WBI was able to engage the participants in understanding and exploring their interests and application areas of the relevant technologies. Breaking into smaller groups, synergisms were identified and opportunities and possibilities for collaboration were developed. The Deep Dive workshop produced several strategic plans for specific technology developments and starting points for further discussions on collaborative activities. In this case, a potential CRADA was identified and the AFRL principal developers were able to roadmap a strategy for the collaborative development of this critical Air Force R&D area. While Deep Dives are not the only way to stimulate and catalyze Collaborative Innovation on nationally important R&D areas, they have become a standard way for AFRL to examine its strategy in critical technology areas and engage multiple entities to assist them in this development.

| COLLABORATION INSTITUTES

In certain strategically important technology areas, a Deep Dive can lead to the development of a highly collaborative and formally structured Collaboration Institute. Such was the case with advanced ladar and optical communications technology. Several Deep Dives on the existing efforts underway in government, academic and industrial organizations were conducted by AFRL and WBI and this led led to the creation of a plan to coordinate all of these efforts over the next ten years. It was clear however that simply coordinating these efforts would not yield the breakthroughs that these technologies offered and a more cooperative and structured activity would be required. After several facilitated meetings to discuss what each of the individual research organizations wanted to achieve over the next five years, it became clear that a formal collaborative initiative was not only possible but desirable by a majority of the players. WBI investigated several potential approaches that could be pursued, described and discussed them individually and collectively with the researchers who were interested in collaborating and held a meeting to essentially seal the deal. The overall objective of the collaborative was to dramatically accelerate the state of the art in ladar and optical communications for the mutual benefit of all of the collaborators. The team eventually settled on the formation of a collaborative institute, the Ladar and Optical Communications Institute (LOCI), to be centered in Dayton, Ohio, near the AFRL Sensors Directorate, the University of Dayton and satellite offices of several companies that were conducting significant research in these areas. The terms of agreement, funding approaches, leadership and management approaches and decision making procedures were all developed and agreed to in less than one month. LOCI is housed at the University of Dayton but it is a collaboration between AFRL, two universities and several companies and is still going strong ten years after its inception. It has successfully stimulated and catalyzed the phenomenal progress that has been made in ladar and optical communications and has spun off several other collaborations in related areas. In this case, collaboration not only led to accelerated progress in technology but also produced a benchmark for innovation in collaborative organizational behavior that has been used by WBI and other innovation institutes to accomplish similar goals.

▌ COLLABORATION ENVIRONMENTS

In 2000, there were no models with similar purposes and objectives to guide the founders in the creation of WBI. There existed a variety of institutes, but most were connected to non-profits, universities and companies. Many of these were singularly focused on helping the sponsoring organization with either technology development or innovation support. While the major objective of WBI was to assist the Air Force, in particular AFRL, with innovation and collaboration, it had a secondary mission to support innovation and economic development in the community surrounding Wright-Patterson Air Force Base. Funding support was to initially come from AFRL but the long term goal was to support the growth of WBI from a wider variety of sources. Importantly, the Air Force acquisition rules had to be followed and setting up a 501-c-3 non-profit funded by the government had restrictions. In an effort to observe best practices and collect some lessons learned, several unique innovation institutes were examined and visited. While this approach did lead to very useful ideas on how to set up and begin operating WBI as an AFRL Innovation Institute, many other insights were obtained. By far, observing and applying the approaches used at Santa Fe Institute in Santa Fe, New Mexico had a major impact on the future of WBI.

In 2001, the Santa Fe Institute was a non-profit organization focused on tapping into the best minds on the planet in order to solve the toughest problems in the world. Located overlooking the beautiful Santo de Cristo Mountains in northern New Mexico, it was designed to catalyze and facilitate collaboration and innovation and it employed very creative approaches to achieve its goals. After several visits and great help from the Institute's staff, it was clear that WBI not only could but probably had to use some of the Santa Fe Institute's techniques and approaches if it was to achieve its mission.

On one visit, we observed a Collaborative Innovation project of ten Nobel Laureates from different fields working on several problems which could only be described as very big and very audacious, for example, the meaning of life, the birth of the universe, etc. While these Nobel Laureates were some of the smartest people on earth, they were from different fields, talked and thought differently, were very independent, intellectually competitive and sometimes a little egotistical. The Institute's challenge was to get them to talk, to listen, to cooperate, to collaborate and to stay focus on the big problems.

The Santa Fe Institute employed a number of practices in order to accomplish these goals and some, while in retrospect not seeming like innovation breakthroughs, were fascinating and definitely worth copying. Just to get the Nobel Laureates to talk was a challenge, so having mandatory coffee'n'donuts and wine'n'cheese sessions for a full thirty minutes every morning and afternoon (respectively) in the courtyard really helped. Sometimes the collaboration and cross-talked worked so well that the staff had difficulty in breaking up these sessions and conversations carried on long after the food and drinks ran out. Since they all had their individual offices and often worked on the challenges independently, a simple way to offer, respond and add to individual ideas had to be develop. This was accomplished by making the walls separating the rooms out of glass and fostering a fascinating communications approach that entailed writing ideas on the glass and being open to adding, commenting and occasionally correcting each other's ideas. A third approach to foster rapid communications and collaborative thinking and innovation was to make all the furniture, and even the walls, easily movable and reconfigurable. This allowed the rapid formation of teams and structural approaches which accommodated the tasks at hand.

The Santa Fe Institute used many creative approaches to foster collaborative innovation and we unabashedly stole as many as we could. The first WBI rooms, which we now call Innovation Environments, employed white-boards to simulate the glass walls, utilized modular, mobile and very comfortable furniture and were equipped with the latest technology to rapidly record, project and manage the ideas coming from the workshop participants. A facilitator was available and generally used in each workshop and mandatory fun-breaks and plenty of snacks and drinks were the rule. While this is standard practice today, it was relatively unique in 2000 and AFRL researchers who worked alone and in a very guarded, sterile environment inside the fence were extremely happy to come to WBI for collaboration and innovation.

| COLLABORATION TOOLS

As we set out on our journey to enhance collaboration in AFRL, we searched for any tools and techniques that organizations and teams had used successfully to achieve collaboration. Given the increasing emphasis that collaboration has received in management and business circles, we assumed that the most recent books, articles and journals would provide us with what we needed. Surprisingly, the most useful ideas were not recent innovations, the three that we found most impactful have been around for a very long time. We named them the three C's: Conscious Conversations, Circle Clusters and Creative Convergence.

There's a small jewelry shop in Santa Fe, New Mexico whose owner,

Ross LewAllen, is both an artist and a Shaman. Ross has traveled the world and has spent significant time living in a wide variety of ancient cultures. While each culture had very specific ways of dealing with its own situational challenges, many employed a similar process when collaboration was needed. In decision-making meetings, an object was passed from person to person and two simple rules were obeyed. The person holding the object had to "speak from the soul" and all others had to "hear from the heart". In the vernacular, the rules came down to "telling it like it is" and "focused listening". The object could be anything but often was an elaborate carved stick that provide a unique way of centering the participants on the importance of the dialog. This process is often called "Conscious Conversations" because it gets to the heart of the matter very quickly. We thought that this would be very compatible with the integrity norm in AFRL and have employed this practice for the past twenty years when important collaborative agreement was needed and other approaches were failing. We have our own "talking stick" for such occasions, actually carved for us by Ross LewAllen. It has been so successful that our "talking stick" has been used to reach important agreements in major corporations, high- level government negotiations and even to set the stage for international peace agreements.

While it might be a myth, King Arthur used an interesting approach to get his knights to agree on how to conquer their enemies and search for the Holy Grail. In the stories, Arthur had them sit at a round table so as not to promote anyone, even himself, as more important than the others. As they sat, they placed their swords in front of them, all pointing to the center as a symbol of common purpose. At least for King Arthur and the Knights of the Round Table, it allowed them to reach consensus and form the powerful collaboration that were needed to achieve their goals. We have employed a similar process at WBI when collaboration between unequal and divergent groups was needed. We literally have them sit at a round table and facilitate the discussion to reach agreement. This is particularly interesting for a military organization like AFRL since we often have a general sitting next to a sergeant with civilians and outsiders inter-dispersed around the circle. While initially somewhat uncomfortable, the participants are quick to see the point and very open and sincere discussions follow. The idea that all are equal and collaboration is the solution quickly becomes apparent. This practice has been so successful that almost all of our workshops include "Circle Clusters" to keep the discussion vibrant and headed to consensus solutions.

Since WBI is an Innovation Institute, one of our heroes is Albert Einstein. We even have a room named Einstein in our Innovation Center and a dedicated library of books focusing on him and his work. In one of them, he is quoted as saying, "No problem can be solved from the same consciousness that created it". Collaboration is generally a very tough challenge, so we often use the Einstein principle of "Creative Convergence" to develop collaborative solutions for difficult challenges. The concept is to look at the problem from multiple perspectives and different consciousness in order to arrive at a novel solution that allows collaboration. We do this by asking individuals with very divergent perspectives to participate in a dialog that focuses on a problem that is common to all of them. Using various facilitation approaches, their perspectives are shared and common themes are developed. In the end, solutions are generated through consensus and invariably a collaboration is formed to further develop and implement the solution. The driving forces of a common challenge, shared perspectives and integrated ideas often produces the convergence upon which collaboration can take place.

| COLLABORATORIES

When we first started WBI, we looked for best practices in government, industry and academia to tailor and apply to AFRL. We found one approach that seemed to fit the bill and decided to build a "Collaboratory" at WBI and use it to help AFRL. As the name implies Collaboratory is a combination of two words, Collaboration and Laboratory. For the environment, we used our findings from other organizations that supported collaboration. The walls of the Collaboratory were either windows or special whiteboard wallpaper that allowed instantaneous capturing of the conversation with erasable markers. The furniture was modular, flexible and easily movable such that reconfiguration into any structure could be done in minutes. The recording, video and projection systems were all state of the art and allowed essentially instantaneous interaction throughout the Collaboratory as well as with other remote site. We wanted no wires on the floor so that participants and furniture were free to move, so we constructed poles that were connected to ceiling power cords with bottom outlets that could be positioned anywhere in the room.

▌ INNOVATION AND COLLABORATION WORKSHOPS

As we investigated organizations that had seriously tried to enhance collaboration and innovation in their workforce, one thing stood out. Every one of them had developed and implemented a broad education and training program that was designed to change the culture in the organization. Leadership was firmly behind the program and often made attendance mandatory for much of the workforce. While it often took a year or more to begin to see measurable results from these efforts, the culture slowly changed to one of Collaborative Innovation. Based on these findings, AFRL asked WBI in 2005 to develop a innovation and collaboration program which could be delivered to a significant percentage of its workforce in a two year period. WBI spent three months developing the program, tested it with several AFRL groups and then executed forty Innovation and Collaboration Workshops to over a thousand AFRL personnel.

The one week workshops were interactive and experiential and kicked off by the AFRL Commander. The fundamental concept was that Collaboration on Challenging problems could lead to Innovative solutions. Each workshop had between 20-25 participants with a wide variety of skills and from organizations across all of AFRL. Some of the workshop modules were delivered to the entire cohort but breakout groups of between 5-7 participants were frequent in order to take advantage of the different participant's perspectives, expertise and experiences. The principles and processes associated with collaboration and innovation, using best practices and lessons learned from across industry, academia and government, were discussed and practiced on real challenges facing the Air Force at that time. We quickly decided that a similar challenge for all workshops would be interesting since we could compare the various cohorts approach to collaboration and innovation. We chose the CBRNE (Chemical Biological Radiological Nuclear and Explosive) threat because of its relevance and importance to national security. On the fourth day of the weeklong workshop, the cohort went into a two-day rapid prototyping mode to design and actually produce a counter to one or more of the CBRNE threats. The prototypes were quite creative and the participants were extremely clever in obtaining the materials needed to construct their designs. The approaches were briefed and the prototypes were demonstrated to the AFRL Commander on the last day. While this was not the objective of the workshop, some of the designs and prototypes were further examined for potential military application by appropriate organizations in AFRL.

▌ COLLABORATIVE INNOVATION SPRINTS

Little did we realize in 2003 that the Innovation and Collaboration Workshop's two-day rapid design and prototyping exercise was a predecessor to what would later be labeled a "Design Sprint". Today, the intense and full-time dedication of a variety of experts on a specific problem over a multi-day period resulting in an early prototype of a solution is called a Design Sprint. While there is some debate on who discovered the process, Google has published extensively on its value and now teaches courses on how to conduct a successful Design Sprint. While it was compressed to two days, our workshop sprints had the same basic elements as the current ones. A structured problem decomposition process was applied to allow the participants to focus on some aspect of the larger problem. Once consensus was reached on the "real" problem to be attacked, a variety of methods were used to obtain as much information as possible, both within and outside of AFRL, about various approaches that might lead to a solution. While the research intelligence algorithms and capabilities of today were not available at that time, the participants used on line searches and library connections to rapidly get the information and insights that they needed. Once they had studied the problem, one or more prototypes were designed and built. Often, these prototypes were very basic and only utilized to test the viability of the solution. Failure was not criticized but used to improve the solutions until a final prototype could be built. Invariably, the prototype was tested, usually after hours and with the help of some of their friends and coworkers. On the second day, further refinements were carried out and tested and a short briefing was prepared to explain the process and the approach that the team had used to solve the problem. The briefing and the demonstration were the final products of the workshop, much like in today's modern design sprints.

CORPORATE DEVELOPMENT OFFICERS

Early in its formation, the leaders of AFRL realized that one of the biggest challenges it faced was that the ten stove-piped Technical Directorates might not collaborate with each other. It didn't take long after the stand-up of AFRL in 1998 to recognize that this would be a big problem. While AFRL is a military organization, it is more a bureaucratic R&D organization that is not driven from the top. So the S&Es in the Technical Directorates don't have to collaborate with each other unless they see some gain in doing so. Unfortunately, the leaders and managers of the Directorates not only didn't see any significant upside, they actually saw it as a potential loss in funding and influence. Many of the Directorates get funding from outside agencies and they thought that collaboration might cost them some influence with their funding customers. The majority of the Directorate funding comes through AFRL itself but most of it is earmarked by Congress or higher headquarters for particular programs in the Directorates. Since there is very little discretionary money available to the AFRL Commander, there was no financial incentives to reward collaboration. While not specifically focused on inter-Directorate collaboration, WBI attempted to foster this collaboration through the Corporate Development Officers (CDOs) that had been created early in the formation of AFRL. Leadership of the CDOs started with Wendy Campbell and later was taken up by Dale Wissman and Cheryl Reed. The primary mechanism used was to have the CDOs heavily involved in the creation, design and delivery of the Innovation and Collaboration Workshops. From the onset, the CDOs were very instrumental in making the workshops focus on the potential benefits of collaboration in AFRL. All the CDOs were involved in the design of the workshops and on a rotational basis, three CDOs helped facilitate each workshop. Throughout the exercises and particularly in the rapid prototyping finale, the CDOs were constantly promoting collaboration and reaching back to their separate Directorates to encourage support to the workshop participants. After all of the workshops were delivered, it was clear that the CDOs were the backbone for collaboration in AFRL. While some of their leaders were not as enthusiastic, the CDOs have strongly supported WBI's collaboration programs and have played a major role in fostering collaboration in AFRL.

| WBI'S INNOVATION AND COLLABORATION CENTER

In 2006, WBI stood up its Innovation and Collaboration Center at 5200 Springfield Street in Dayton, Ohio. At first, we called it the "Wright Brothers Institute Innovation and Collaboration Center" but this was too long a name and the acronym WBIICC was not easy to pronounce. Evidently the name was important enough to command the attention of the Commander and Executive Director of AFRL and they strongly suggested that we call it "Tec^Edge". The idea was that our new facility was not only on the edge of Wright Patterson Air Force Base but that it was also working collaboration and innovation at the leading edge of technology.

The original Tec^Edge was on the second floor of 5200 Springfield and occupied about 10,000 square feet. The center included the Collaboratory where all the Innovation and Collaboration Workshops were conducted and various meetings to promote collaboration in AFRL and the Dayton Region were held. There were four breakout rooms, a decommissioned Skiff that housed our electronic systems, a café and make-shift exercise room, several casual gathering spaces and a couple of offices to house the staff. The walls were covered with white boards and there were shelves to display some of the prototypes that had been built in the workshops. While not elaborate, the furnishing and wall décor was specifically chosen to promote innovation and collaboration, as evidenced by highly convertible modular desks and a clock that ran backwards. It was small but it was the beginning of what we had taken on as our purpose, collaboration and innovation both inside and outside the fence. In 2007, a new Tec^Edge was created at 5000 Springfield Street across the parking lot. This one had 25,000 square of space and we took advantage of all the experimentation that we had tried over the past five years to truly live up to its real name: WBI's Innovation and Collaboration Center.

| HALO

One of the many approaches that was tried to improve collaboration in AFRL was the HALO room. In 2005, WBI was asked by Dr. Alok Das to create a virtual electronic collaboration system for AFRL utilizing the latest conferencing technology. We contracted with Hewlett Packard to build a high-end HALO collaboration room at Tec^Edge that was connected to similar rooms at Eglin AFB, Kirtland AFB and Griffiss AFB. With that coverage, all of the key AFRL Directorates could participate in near-real conversations, meetings and hopefully collaborations instantaneously. While expensive at $1 Million, the fidelity was so good and the latency so minor that people in all of the HALOs felt that they were in the same room. Nuances, like a smile, glance or frown were obvious and collaboration conversations between personnel from all of the AFRL sites were enhanced using the HALO capabilities. One lesson that we learned from the HALO experiences was that uninterrupted, real-time, face-to-face conversations always produced better results than those that utilized any of the other communication technologies available at the time. The HALO rooms at WBI and the other AFRL sites were upgraded and utilized for several years until collaboration technologies caught up. Today, WBI utilizes Zoom Rooms (at a fraction of the HALO costs) for much of its conferencing and communications services and is even developing distributed and virtual workshop facilitation capabilities to extend its innovation and collaboration support services.

▌DISCOVERY LAB

To demonstrate the possibilities of collaboration and innovation, WBI worked with Dr. Rob Williams of the AFRL Sensors Directorate to create a Discovery Lab for high-end innovation in the virtual world. Based somewhat on the Second Life virtual experience, the Discovery Lab allowed participants to create an avatar of themselves and enter a virtual world to collaborate and innovate with other avatars. Initially, the virtual world experience was experimental and served to acclimate and educate students to the possibilities of operating and collaborating in a virtual world. As the capabilities of the Discovery Lab matured, Dr. Williams and WBI collaborated on summer long projects involving students from a variety of universities to collaborate and innovate in a virtual world. After several years of successful operation, we labeled the program "Summer at (Tec^) Edge" or SATE and about one hundred students went through the program each summer. The students entered with no special talents or computational capabilities, were trained to enter the virtual world as avatars and to operate in that world to collaborate and solve difficult problems, all virtually. Lectures by relevant experts were delivered to the student-avatars, workshops were held, designs were created and prototypes were built and tested, all in a virtual world that replicated the physical space at Tec^Edge. The experience for the students was so profound that many of them changed their majors to continue working on their SATE experiences and nearly all were positively affected by the program. After a few years, some of the students wanted to continue this experience and a "Year at the Edge" or YATE program was developed. Since the workspace was virtual, students and professors could interact from different locations and collaborate in a unique way on problems of mutual interest. SATE and YATE are wonderful examples of how collaboration can work using novel environments and advanced technology tools. Uniquely, most of the successful collaborations that were achieved occurred between students, most of them in their early twenties who had grown up being comfortable in the virtual world of games. The idea of collaborating and innovating in a virtual world never took hold with the more senior R&D community within AFRL. Fortunately, Dr. Williams has continued to pioneer this area and is hopeful that it will become a powerful way to achieve collaboration.

INNOVATION

"The Wright Brothers flew right through the smokescreen of impossibility", Charles Kettering

▌THE IDEA LAB

In 2008, we decided that it was time to do something serious about innovation, so we created the IDEA Lab. At first, it was just a concept. The title "IDEA" was an acronym for Innovation Development, Experimentation and Application and we added "Lab" to indicate it was a place where we could learn and experiment with various techniques, approaches and tools to enhance innovation. While the IDEA Lab did occupy a number of places between 2008 and 2018, it was much more than a place, it was a way of thinking and being. Our first IDEA Lab experiment was a problem solving session in a hallway of Building 5100 with a number of students that we had recruited during the summer of 2008. Later we went outside to the parking lot and made these sessions more fun, often ending with eating, drinking and game playing. When we moved into Building 5000, we designated an odd-shaped space in between collaboration rooms as the IDEA Lab and set up desks for five professionals and five student interns. When that space was needed to create a conference room to hold 60 participants, the IDEA Lab was moved to the back of the building and we created a no-walls, group environment with sofas and lounge chairs and lots of crazy creations that was always noisy, well supplied with food and beverages and focused on learning and having fun. In 2009, we hired Emily Riley, an innovation practitioner from P&G and 4inno to run the IDEA Lab and when she left in 2012 to head up the innovation program at General Mills, Cheryl Reed, the former lead CDO from AFRL, became her successor. In 2019, the IDEA Lab moved to the second floor of Building 5000 and was transformed into the Innovation Services Division of WBI under Steve Fennessey.

The IDEA Lab concept was to develop, modify or apply any interesting innovation tool, technique or process to helping AFRL teams with innovation. We started with creative problem solving and other well-known innovation approaches that had been around for many years. While these certainly helped, particularly to give teams a process to follow, we were always looking for more powerful and creative approaches. During the first year, we searched throughout the world for best practices and seriously studied novel concepts that were being used by companies and universities in the United States, Europe and Asia. To our surprise, we found some of the best just down the road at Procter and Gamble in Cincinnati, Ohio. P&G had instituted a company-wide program called Connect and Develop that according to the Harvard Business Review, had produced outstanding results. Because it was so close, we spent many days visiting P&G and their consulting partners to learn everything we might be able to apply to the Air Force. Even though they are a consumer products company, we found that much of their processes and techniques were applicable to a military research and development organization like AFRL. Over a period of two years, the IDEA Lab modified the P&G ideas for AFRL use, joined in several partnerships with P&G and their support contractors and created a baseline IDEA Lab process which WBI has used ever since with AFRL and other customers. WBI set up the IDEA Lab organization similar to that being implemented in most of the country's leading consumer products and R&D organizations, with Front End and Back End of Innovation hubs and very substantial utilization of research intelligence, global landscaping, multi-site communication hubs and the latest collaborative information-sharing technology. While the IDEA Lab started as a WBI experiment, it soon became a way to significantly enhance innovation in AFRL. At first, AFRL teams came to the IDEA Lab to get help with facilitating their problem solving or program planning workshops. When the results from their these workshops were presented to AFRL management, the leaders of AFRL quickly saw the benefits of using IDEA Lab tools and techniques and made a point of sending any team focusing on innovation to the WBI

IDEA Lab to get help. As the use and benefits of the IDEA Lab increased, we decided to create a mission statement for the IDEA Lab. It read: "The IDEA Lab's mission is to significantly enhance the speed, creativity and return on investment of AFRL by generating new insights into complex challenges, uncovering key areas for investment, and quickly connecting with new partner, problem solvers, ideas and solution in the USA and around the world." What started as an experiment soon became a major enterprise of WBI and a force-multiplier capability for the Air Force Research Laboratory.

| IDEA LAB INNOVATION MODEL

The basic IDEA Lab innovation model used from 2008 to 2018 can be described as a series of Front-End of Innovation (FEI) steps connected to a broad array of Back-End of Innovation (BEI) opportunities. In practice, it's actually simpler than that. Following Einstein's edict, we started with the problem space, stayed in it as long as needed, transitioned to possibilities and opportunities and then moved into the solution space. The key was to stay in the problem space for as long as possible, making sure that new insights were incorporated into the challenge or opportunity, redefining the problem if necessary and, only then, transitioning to the solution space and looking for new solutions, opportunities and partnerships. We quickly realized that this was difficult for our AFRL partners to do. Since most of AFRL's personnel are scientists, engineers and financial managers, they generally were much more interested in coming up with solutions than staying in the problem space. Since many had been around for a long time, they were eager to offer a quick solution or to dismiss potential solutions that they had tried and just didn't work. That is the challenge of staying in the problem space, but we were as strict as we could be and despite protests, it generally paid off in the end. Once we reached an agreed-to redefined problem, we then focused on coming up with opportunities and possibilities that might lead to solutions. In the solution space, we tried to encourage the teams to try new approaches, like rapid prototyping, minimal viable products, and crowdsourcing using global innovation challenges. While these approaches seem commonplace today, this was not the case in 2008 and new paradigms had to be developed, taught and practiced to take advantage of these new innovation techniques. While the basic problem space/solution space model made sense to us, we filled it in with more specific processes and applications so that the AFRL S&Es would see how it could be used to help them in their work. While each process or application will be described in more detail later, they all fit into this simple model that allowed us to explain our approach in a way that was appealing and resonant to the science and engineering foundations and culture of AFRL.

▌ AFRL INNOVATION WORKSHOPS

Between 2008 and 2018, we conducted two to three workshops per week for AFRL teams. That's about 1000 workshops over that ten year period, so we certainly were busy with applying our process and helping AFRL in any way that we could. But throughout this time, we also experimented with new approaches that we either heard about from colleagues, at innovation conferences or developed ourselves. Some of these will be discussed later, but the experimental nature of the IDEA Lab was very refreshing to those of us at WBI. Every workshop started with a planned process, but if we thought that there might be better results by trying something new or modifying our process, we did so. This was sometimes difficult for the more structured members of our team or participants in the workshop to accommodate, but it generally worked. After each workshop, we hot-washed the workshop and expanded our lessons learned and best practices file for possible process modifications in future workshops. The IDEA Lab was truly a learning organization and that allowed us to attract some of the best practitioners and facilitators to WBI for employment. Of course, some turnover occurred because many of our employees were sought after by other companies interested in innovation and most were extremely successful in their future endeavors.

The nature of the AFRL workshops ranged from problem-solving to solution seeking, tactical to strategic and broad to specific. We often were engaged by S&E teams who had tried and failed many times at solving their problem and we coined the phrase, "When you're stuck, ask us". Some of their problems were very specific and we used research intelligence and global landscaping to see if there were solutions already out there. If not, we created possibilities and brought in outside experts to enhance the AFRL teams. It was not unusual for the initial workshop to morph into several more specific workshops once the problem was redefined and fractionated into several sub-problems. Every workshop was preceded by several planning meetings between the AFRL problem owner(s) and the IDEA Lab team in order to maximize the efficiency of the workshop, get agreement on the objective of the workshop, identify who should be at the workshop and finalize all the logistics required.

To help new IDEA Lab team members with conducting an innovation workshop, a Workshop Guide 101 was created which outlined twelve pre-workshop actions that should enhance the probability of the innovation workshop being successful.

1. Get to know the workshop owner as much as possible. Try to understand how this particular project and workshop fits into their short and long term plans. If possible, try to understand their style since it will really help during the workshop.
2. Work with the problem owner to get a very clear picture of the objective of the workshop. If necessary, challenge it if you think it's not really what they want. Iterate with them until you both agree that this is the correct objective and it is feasible for the workshop to achieve this objective.
3. Develop an agenda for the workshop with the client. Even if it's not detailed and there are some uncertainties, get agreement with the client on an agenda.
4. Determine who will be at the workshop and identify every attendee's role and every attendee's responsibilities during the workshop. The client's role is always the owner of the workshop and your role is always the facilitator.

5. With the above, create an OARR (Objective, Agenda, Roles, and Responsibilities) document and have the client send it out to all workshop attendees prior to the workshop.

6. Do everything you can to physically prepare for the workshop, particularly with logistics, food, recording, IT systems, etc. You should try to get as much help from the client on this so as not to burden yourself with these issues during the workshop. If the client cannot accommodate some of these, then they will need to be supported by someone in WBI or an outside contractor.

7. The physical layout of the workshop room is important. Basically, if the workshop will be a more formal event, use a U-shape layout. If the workshop will be interactive, collaborative and not hierarchical, use clusters. Clusters is a great default layout because it assures communications and you can always move to more structure by assigning people to the clusters

8. Have name tags for every participant. Big first name with smaller last name underneath usually works. No rank or organizational symbol is preferable. You can always add name tents if needed.

9. Always bring large (8x10) post-it notes and black markers for planned use or just in case. It's a great way to have teams produce inputs and then stick them on the whiteboards to report out (so much better than writing on the boards or using easels).

10. Start the workshop on time by introducing yourself and WBI. Hand out the OARR document before or at the start of the workshop. Before introducing the owner, ask if there are any questions about anything. Only then, turn it over to the owner who should start with the why and the objective of the workshop.

11. The OARR document is the key to facilitating the workshop. Basically, your primary job is to keep the workshop focused on the Objective, maintain the timing, flow and activities described in the Agenda, use the Roles to identify who should do what and when during the workshop and constantly interact to make sure all the participants fulfill their responsibilities.

12. Sometimes it's difficult for everyone to stay in their role, particularly the owner. So watch that person carefully. However,

watch yourself as well because it's also very easy for you to take on someone else's role, like recording, or leading the workshop, or actually doing their work for them. Innovation is difficult and the participants might be tempted to ask you to do their job.

Because AFRL receives billions of dollars of basic research and exploratory development funding for its programs, one of its primary missions is to advise the Air Force on long-term technology possibilities, opportunities and threats. Many of the workshops were really very strategic investment planning workshops and required the most innovative thinking possible giving the consequences of missing an important technology opportunity or neglecting how the combination of several technologies could make a game-changing difference to the Air Force of the future. These strategic planning workshops were particularly challenging since the team would always start with an extrapolated version of the current approach and the innovation came from moving the team to new ideas and possibilities. In some cases, a paradigm shift from the past was needed and often that did not come easy. In many cases, the IDEA Lab facilitators had to continually challenge the participants to think differently and to look at their problems and approaches from a very different perspective. Many times, Einstein came to the rescue with his quotation, "No problem can be solved from the same level of consciousness that created it".

While you might assume that AFRL focused only on technological innovation, many of the workshops were attempts to find new ways to accomplish the myriad of business activities that a large bureaucratic organization requires. The objectives of many AFRL innovation workshops were to create new financial management strategies and processes, to find ways to attract, recruit and retain high-performing individuals, to accelerate the development of streamlined acquisition process and to reduce the difficulty of using information technology while maintaining the security required by a federal military agency. While the technical innovation workshops were the bulk of our activity, the IDEA Lab's innovation workshops produced some very creative and powerful business processes that revolutionized the way that AFRL did business.

▌COMMANDER'S CHALLENGES

One of the best applications of the complete IDEA Lab innovation process occurred with the AFRL Commander's Challenge teams. In 2008, the AFRL Commander instituted a program to have volunteers from the AFRL S&E corps form teams that would compete in solving real-time, important military problems. These problems ranged from stopping an attacker from entering a building or a gate at an Air Force Base to downing an ultra-light airplane that might be carrying contraband across a border but without hurting the pilot. Over a period of ten years, there were twenty such challenges that involved several hundred S&Es across the nation. The team members had to leave their current positions, form a dedicated team and devote about six months to the challenge. Because this often meant not returning to the same position after the challenge was over, the Commander's Challenges attracted mostly junior military and civilian S&Es who were eager to more directly contribute their engineering and scientific talents to the mission of the Air Force. Some of the teams located in Dayton and took advantage of the WBI innovation and collaboration facilities to work the challenges.

In many cases, the teams were self-organized and consisted of 5-6 junior civilian or military S&Es who had never met before and came from various AFRL Technical Directorates. Their first meeting often occurred at WBI and was facilitated by IDEA Lab personnel. Several days were spent getting to know the background, experience and expertise of the team members and organizing the team for the six month activity. We helped them with various techniques like the Myers Briggs behavioral style instrument and other tools that identified the talents, perspectives and unique characteristics of each member. It was fascinating to watch six people who had never met before create a team that took advantage of the strengths of each member to maximize the effectiveness of the team. Generally, problem solving teams are pre-formed based on the participant's job titles and years of experience and the leaders of these teams are assigned by upper management. In this case, it was more like a scratch sports team that iterated their responsibilities until the potential of the team was optimized. In essence, the team was a collaboration of willing members who wanted to be there and were totally focused on solving the challenge problem. While tiger teams and problem-focused teams do get stood-up in most organizations, the unique way that these teams were formed seemed optimal for developing Collaborative Innovation solutions to the challenge. And that is exactly what happened.

Once the team was formed and organized, IDEA Lab personnel helped the team examine the challenge using the front-end of the IDEA Lab process. As predicted, the team wanted to generate solutions immediately and, while we captured initial solution ideas for future use, we pushed back and kept them in the problem space for as long as possible. Eventually, all team members saw the value of staying in the problem space and we often were able to keep them there for as much as a month, sometimes two. We used a variety of techniques to break down the problem with the most interesting being functional decomposition, which will be described in detail later. Functional decomposition takes you to very different sub-problems and the team sometimes created sub-teams that would investigate these more specific aspects of the challenge. Global landscaping and research intelligence was seriously applied to all aspects of the problem and the IDEA Lab professionals provided their resources, expertise and intelligence capabilities to the team. Several weeks were specifically dedicated to examining and understanding anything in the world that was related to any aspect of the problem. This global landscaping capability was enhanced over the years and what began as literature searches using library data bases in 2008 grew to highly sophisticated machine-learning integrated intelligence software systems in 2018.

One of the breakthroughs that occurs when a team uses functional analysis to redefine a problem and then applies extensive research intelligence to find expertise in the various sub-problem area is the identification of individuals that might provide a new perspective on the problem. Once identified, these individuals can be engaged by the team and very new approaches to solving the problem can be developed. In particular, the functional breakdown of the problem often leads to functional expertise that would never have been sought because it is used to drive solutions in very different problem spaces. But this unique expertise can be very useful in stimulating new and novel ideas to the current problem. This is particularly true when teams have great knowledge and experience in areas that they believe apply to the problem and only look to their research colleagues in that area for solutions. New perspectives from experts that have never been engaged with the team's problem often lead to creative solutions which develop into innovation opportunities. Most challenges being encountered by our military forces today require multi-disciplinary solutions and having expertise or even ideas in only one discipline will not lead to innovative solutions. One of the powerful outcomes of the initial Commander's Challenges workshops was the realization that divergent perspectives and contributions can lead to unique innovation. The result was for the IDEA Lab to investigate other organizations and opportunities where divergent thinking was utilized for collaborative innovation. This led to the development of Divergent Collaboration workshops which we have refined and practiced over the past eight years. Divergent Collaboration is such a profound technique that it is now an important part of our innovation tool box and is often used in the Problem Space process as well as a stand-alone offering to teams that truly want to open up both the problem and solution space to very different possibilities.

Another tool that is part of the IDEA Lab innovation process is Connect and Explore. This will be discussed in more depth later but its primary focus is to connect the problem owner with the team or individual that generated the problem and will be involved in testing and using any solution that comes from the innovation workshop. The details of how the owner and user are utilized in a Connect and Explore process are very important but the primary intention is to have the user or customer intimately involved in the problem definition and redefinition as well as the solution development, testing and application process. While getting the user involved sounds like a logical approach, how it is done, when the user gets involved and when to bring in the boundary conditions that the user has to currently meet are critical items and will be discussed later. Over the years, having the user as an integral part of an innovation process has gained much favor and it is now common for users to be present throughout an innovation workshop or sprint. The lessons learned in our earlier workshop provide great lessons and insights for how this is now done.

Once the problem has be redefined, we move into the solution space. The IDEA Lab has used a variety of techniques to help in the solution space, many of which have been used in creative problem solving over the past decades. A wide variety of solution generation techniques, like brainstorming, synergism and many of deBono's idea generation techniques are usually applied. To all of those, the IDEA Lab added three additional approaches: in-house or contracted research to further test and develop an idea, rapid prototyping and global innovation challenges.

While all innovation workshops did not have the resources to conduct serious research on potential opportunities or solutions, the Commander's Challenge teams had the time and a significant amount of funds to do so. After one or two months, these teams, collaborating with other S&Es in AFRL as well as AFRL's in-house contractors, began to design and build potential solutions experiments that gave them tremendous insight into what might or might not work for the challenge. Often, the test articles or processes were very early prototypes, designed to yield information or knowledge even if they were destroyed during testing. Sometimes, the teams could piggyback on existing research being conducted in AFRL and occasionally used flight and space testing to understand a potential solution. The lesson that we all learned is that solution development and application is always more probable if the innovation process can include design, development and testing by the team. Ending a workshop with a few good ideas and hoping someone picks them up and proves them out never works.

While prototyping is certainly one way of continuing research on potential solutions, prototyping is much more powerful if it is purposely included in the process from the very beginning. If a workshop has set aside time, resources and expertise to actually build one or more potential solution prototypes, it changes the entire nature of the innovation process. For the Commander's Challenge teams, committing to prototyping as a part of their challenge was natural since they had the longer schedule and funding to do it. However, even the commitment to produce a very inexpensive, rapidly developed prototype in a one-week workshop changes the nature of the activity. As we introduced this concept to teams that had a very limited schedule and funding commitment, the workshops became much more successful. Eventually, we included rapid prototyping as part of our standard innovation process whenever we could and it really paid off. Again, the Sprint process that is now used extensively to produce innovative solutions in many organizations regularly includes rapid prototyping, testing and iterating before the end of the Sprint.

The third solution space technique was global innovation challenges. Initially, for any problem area, the IDEA Lab explored ways to crowdsource the problem and this often led to the identification of individuals, teams and organizations that were willing to collaborate with us. Sometimes, interesting contributions came from this crowdsourcing and occasionally an actual solution, either as a design or a prototype. We began to explore a number of companies who offered a process to scan the world for potential solutions to the challenge or even a modified version of the challenge. Since some of the AFRL problems could not be exposed to the world in their original form, we began to generate unclassified versions of these challenges and used a variety of web-based systems to obtain solutions. Some of these endeavors were more successful than others, but we always included and offered this tool as part of the solution space process. By getting an up-front agreement to potentially use global innovation challenges if appropriate, the nature of the innovation workshop changed significantly. With the Commander's Challenge teams it was always an option and occasionally used. In any case, the overall process led to solutions, solution opportunities and potential partnerships to develop a solution.

THE PROBLEM SPACE

One of the areas of innovation that we have emphasized at WBI is the "Problem Space". While it is an Einstein axiom, staying in the problem space really does pay off. Regardless of whether a team is interested in solving a problem, developing a solution, researching an area or laying out a plan or roadmap to achieve a goal, we spend an inordinate amount of time in the problem space. There are two fundamental reasons for doing this. First, the more you know about the problem, the easier it is to come up with a solution that really works. As all aspects of a problem are teased out of the problem owner or team, insights can be gained on where to look for solutions or who to add to the team to help with the problem. Often, the person who initially defines the problem or challenge does so from his or her perspective and actually limits the problem space to his or her discipline. By turning the problem over and under several times, new viewpoints are generated and the need for multi-disciplinary and functional analysis becomes clear.

One simple but interesting technique to generate new insights into a problem is to decompose the initial problem into a hierarchy of super- or sub-problems. Starting with the initial problem, we ask "why is this a problem?" and that leads to a super-problem that if solved would likely solve the initial problem. By asking why a second time, an even larger super-problem is generated which may or may not be useful to consider. Obviously this can go on again but two why questions usually do the trick.

On the other end of the spectrum, asking "what's stopping you from solving the problem?" reveals more details about the problem. If one of these sub-problems is really what the problem owner want to solve, then the process can focus on that. Again, a second round can occur and even more details about the initial problem are revealed. Regardless of whether a higher or lower version of the problem is chosen or the team returns to the original problem, the exercise keeps you in the problem space and provides many interesting perspectives and insights on the problem.

The second reason comes from our experience helping hundreds of teams with challenges and problems. Unfortunately, many teams work very hard solving the wrong problem. Often they come to us after all this effort and are willing to take a better look at the problem. And sometimes it takes the above problem analysis to open up the problem space and identify the real problem that needs to be solved. The reason this happens is because it's natural for people, particularly highly educated experts to always look to their discipline for solutions to their problems. So they define the problem from their particular perspective, look to their discipline for solutions and unfortunately end up solving the wrong problem. If nothing else, a lot of time and effort can be saved by simply dwelling for a few minutes on whether the stated problem is really the one we want to solve.

▌ FUNCTIONAL PROBLEM ANALYSIS

WBI became acquainted with functional problem analysis thanks to the IDEA Lab's first employee, Larisa Dmitrienko. Larisa had come to Cincinnati from Moscow in the 1970's and brought with her an expert knowledge of a Russian process called the Theory of Inventive Problem Solving or TRIZ. TRIZ is a system for creative problem solving, commonly used in engineering and process management. It follows four basic steps: define the specific problem, find the TRIZ generalized problem that matches it, find the generalized solution that solves the generalized problem and then adapt the generalized solution to solve the specific problem. In order to find the generalized TRIZ problem, the specific problem is examined and analyzed to identify the fundamental functions involved in the problem. Once those functions are identified, TRIZ can be used to determine how problems with that same functional issue have been solved in the past, in particular by examining a long history of patents. As the functional problem and solution is examined, clues to solving the same functional issue in the original problem can be gained.

For example, if the specific problem includes a predominant function like replicating an original process or item a thousand-fold in an efficient and effective manner, similar functional problems and their solutions can be examined for possible solutions to the specific problem. While the TRIZ process is in itself a powerful innovative tool, we used the functional aspect of TRIZ as a principal method of analyzing, separating and re-defining a problem. Functional analysis became an important part of the problem space effort for any WBI problem-solving project. Almost immediately after accepting a problem, we decompose it using functional analysis, identify the key functions involved in the problem, search for problems in any other domain that have that same key function and then look at how that domain solved that functional problem. For example, an important operational problem for the Air Force is how to direct a missile from an Unmanned Air Vehicle (UAV) or drone to a specific target while minimizing the collateral damage to any other part of the target environment. Functionally, this involves a precise understanding the environment and situation and then directing the missile to perform a precision strike of the target. While this was not our first idea, we eventually recognized that robotic surgeons have the same functional problem and have developed techniques, technology, training and operations to develop situational awareness and precision operations in a particular environment. The environment of the robotic surgeon is a particular part of the human body, the situational awareness is a precise knowledge of the current anatomy and the necessary precision generally involves destroying, removing and repairing part of that anatomy. While the scenarios are very different, the functions and the generic solutions are very similar to a drone pilot performing a surgical strike in a complex situational environment. This particular functional analysis was specifically used in one of WBI's Divergent Collaboration workshop with one of the participants being an eminent Ob-Gyn Robotic Surgeon, Dr. Bruce Bernie. As will be described later, his expertise and inputs led to a very unique solution to a critical military problem.

As mentioned, functional problem analysis is used by WBI in many of the projects it takes on for the Air Force. Since the tendency to solve a problem is to look to the experts in that problem space, this approach is quite surprising and even disarming to those experts. Why should a robotic surgeon have ideas on how to improve drone operations in the battleground environment? While it is not the obvious approach, it can lead to very creative possibilities and sometimes very innovative solutions.

▌ GLOBAL LANDSCAPING

One of the most powerful tools of any innovation process is identifying, understanding and applying any data, information, knowledge and wisdom that might be relevant to the problem. While this is often done in the solution space where potential ideas or partnerships might be applied, it is particularly important in the problem space and the front-end of the process. There are many terms for this process but we use global landscaping to describe the search for information and knowledge throughout the world. As WBI matured and high-performance research intelligence technology became available, we described this process as Comprehensive Integrated Intelligence (CII) and that will be described later. From 2005 to 2017, global landscaping was always a part of the innovation process and essentially consisted of scouring all the available information, documents, conference papers, research summaries, books, reports and patents for any information pertinent to the problem. In the early years, this was done mostly through library research. With the advent of major information search companies like LexisNexis and Google as well as information and data bases that could be licensed, global landscaping was greatly expanded. In hours, huge amounts of relevant information could be collected. Further, tools were available to analyze the sources of this information in order to suggest where further information could be obtained. For example, authors from one paper were connected to co-authors from another paper to literally produce a paper-trail where more information could be found. WBI research specialists were on top of any the development in this field and quickly tailored and utilized these advances for the benefit of our customers.

Two specific global landscaping techniques that were extensively used by WBI need mentioning. In 2007, WBI entered a partnership with Procter and Gamble and General Mills to sponsor the development of a comprehensive tool called inno-360. This end-to-end innovation software tool was developed by 4inno, a spin-off of the P&G Connect and Develop innovation program. Inno-360 was developed over a period of several years with significant requirement inputs from P&G, General Mills and WBI and became a useful innovation software tool to assist individuals and groups in AFRL. One of the best features of inno-360 was its global landscaping process which utilized imbedded connections to a large number of research intelligence sources that could be supplemented with specific data sources specifically connected to the problem domain. By using inno-360, AFRL problem owners could go on-line, introduce and decompose the problem and then access hundreds of relevant research data bases that might potentially assist with the problem. This global landscaping feature provided very valuable and often new and novel insights into the problem space and could then be used later in the process to connect to potential problem solvers and solution providers. WBI introduced inno-360 to dozens of AFRL S&Es, trained them in the use of the software and assisted them as they used the tool. Unfortunately, the inno-360 tool was not adopted by a large number of AFRL S&Es, as we had hoped. While it was and still is an effective innovation software system, its on-line nature doesn't provide the in-person support afforded by WBI when AFRL problem owners engage with the IDEA Lab facilitators in a "live" workshop with several planning meetings. This was definitely a factor in AFRL's decision to only sparingly use the inno-360 software.

Another very powerful global landscaping tools that still is a key feature of WBI's IDEA Lab process is Tableau. WBI was introduced to Tableau in 2007 by our IDEA Lab information analyst, Lisa Russell, and we have been using advanced versions of Tableau since then. Tableau is a high-end data software system that combines SOTA computer visualization tools with scalable platforms needed to understand massive amounts of data and information. Tableau is an outstanding way to assimilate, categorize and present huge amounts of information from extensive global landscaping in a way that communicates insights, learnings and knowledge. Most importantly, Tableau allows global landscaping to be presented in a manner that allows deep understanding by the user and rapid decision making for further action.

| CONNECT AND EXPLORE

There is nothing more useful in an innovation challenge than direct involvement of the actual user or operational implementer of the capability. To emphasize this approach, Bob Lee of the WBI IDEA Lab developed a process called Connect and Explore which specifically involved the participation of the user in all phases of the innovation process. While most problem solving approaches incorporate users in the early definition of the problem and in the solution evaluation phase, the WBI Connect and Explore process includes the user in every phase of the activity.

The user of a capability, whether a process, system, technique or tool, knows the current situation better than anyone. The user can set the baseline for the innovation challenge and will have opinions and insights into how the capability can be improved. As the problem is functionally decomposed, the user can interact with the innovation facilitators and guide them on the details of the capability, the approaches for improvement that have been tried and the current and future operational environment of the capability. Often, dialogue with the operational user will identify functions which are not apparent to someone not familiar with the capability and this can lead to the involvement of different experts in the problem space. As global landscaping is undertaken, the user can help the research analysts with keywords, probes, connections and tangential inquiries. Continuously connecting and exploring the problem with the user not only produces better results but often significantly expedites the innovation process. As we used this process, we developed best practices and learned some important lessons.

Continuous involvement of the user requires substantial planning and should drive the timetable of the innovation process. Often problem owners are willing to be involved in part of the innovation process but are too busy to be constantly present. Continuous involvement requires a serious commitment by the user and, without that, the Connect and Explore process should not be used. In Connect and Explore projects, the user is an integral part of all aspects of the problem and solution space and planning and scheduling the details of the process assumes that commitment.

A clear definition of the objectives, agenda details and all of the participants' roles and responsibilities prior to and during a Connect and Explore workshop is a must. The clearer the objectives and the smoother the execution of the agenda, the better the process and the outcome. The roles of the user, particularly with respect to that of the facilitator, should be crystal clear. Often a user gets so involved that he or she wants to run the workshop or move too quickly in the agenda. Similarly, the user provides expertise needed in both the problem space and the solution space. Facilitation should be left to the facilitator who can devote all of his or her attention to the agenda and the focus on the objective.

When it's done well, Connect and Explore workshops can produce powerful results in a minimum time. In one case, a team of ten users interacted with two dozen different experts and potential solution providers over a period of four days in a heavily facilitated workshop with outstanding results. The topic was to minimize the time and maximize the effectiveness of complex decision making in a critical battlefield environment. Five specific solutions, three of which could be implemented in the near term, were identified and passed on for additional analysis, experimentation and prototyping. That challenge had been the focus of considerable attention by the user for several years with very little progress to a solution. With one month of planning and a four day workshop, the user walked away with several near and far term solutions.

▌ DIVERGENT COLLABORATION

Of all the innovation processes developed and implemented by the WBI IDEA Lab, Divergent Collaboration has received the most attention. The concept started when a collaboration expert, Peter Benkendorf, approached WBI with a proposition. Peter had been trying to get a number of artists in the Dayton area connected to scientists and engineers at Wright Patterson Air Force Base in order to explore potential areas of collaboration. He asked the IDEA Lab if we could help and, in the spirit of innovation and collaboration, we agreed to orchestrate a meeting between the artists and some senior engineers from AFRL.

The Dayton artists were an eclectic group of musicians, choreographers, painters, sculptors, dancers, videographers and graphic artists. With their input, we brainstormed who within AFRL might be interested and open to connecting with them. We decided the best AFRL group to ask would be the 711th Human Performance Wing's Cockpit Display team. This group is in charge of developing cockpit interfaces for advanced aircraft and is always investigating new ways to deliver information to the pilot in a more effective and efficient manner. Over the years, this team had developed the technology and the systems that changed aircraft cockpits from a few dials and pressure gauge instruments to the glass cockpits of today's fighter airplanes and the development of Heads-Up displays, Helmet-Mounted displays and Virtual-Imaging displays for high-performance aircraft. We narrowed it down to eight S&Es from the team and invited them to a four hour meeting at WBI to meet and talk to eight of the Dayton artists.

The meeting room was set up in U-shape form with eight seats on each side and room at the top for Peter and several members of the IDEA Lab team. As they entered the room, the S&Es, dressed in suits and ties and carrying computers, all sat on one side of the table. The artists, dressed casually and some carrying examples of their work, sat on the other side of the table. It resembled a Junior High School dance and we really didn't think that the meeting would go well. After introductions, we asked each of the participants to describe their work and some of the challenges that they were facing. That discussion lasted about an hour and then we took a break. When we came back to the room, it looked very different. The engineers had mingled with the artists, the mobile furniture had been reconfigured into clusters and small mixed teams of artists and engineers had self-selected and were having vigorous discussions. And, as they say, the rest is history.

The interaction and interchange of information in the clusters was so intense that we had some difficulty getting the meeting back to order after another hour had passed. The S&Es were showing the artists charts and designs on their computers and the artists were displaying and describing their work to the engineers. When we did bring the group back for overall discussions, they told us that they had made plans to meet outside of this meeting, the artists had been invited into the base to participate in cockpit display briefings and demonstrations and the engineers wanted to see more of the artist's work, including a tour of the Dayton Art Center and a performance of the choreographer's ballet play. The potential collaboration between this really diverse, even divergent, set of individuals was off and running.

After the initial meeting, several sub-groups engaged to work on problems or challenges of mutual interest. In these sub-groups, the engineers provided technologies to enhance the artist's work and the artists helped the engineers look at visual displays in very new and creative ways. As an example, the choreographer produced a ballet of unmanned aerial vehicle (UAV) drones that was filmed and won a local artistic performance award. Another involved the artists adding three-dimensional visual, sound and motion enhancements to UAV ground-pilot stations to enhance the pilot's effectiveness and efficiency in controlling multiple UAV systems. This initial collaboration between very diverse professionals was so successful that WBI decided to use the lessons learned in a new IDEA Lab innovation process which called "Divergent Collaboration". We then expanded our Divergent Collaboration activity to bring in other user groups.

We started with a two-pronged approach, a summer program for students and an experimental workshop to solve an Air Force problem. In the summer program, several students with diverse academic interests were challenged with a problem of relevance to the Air Force. They were given a workplace and laboratory at WBI to meet and work and they were assisted by an IDEA Lab facilitator to guide and assist them with whatever they needed. Although we were hopeful that they would come up with creative solutions to the problem, we were more interested in their collaborative dynamics and problem solving approaches. We were successful on both counts that first summer and the lessons that we learned and best practices that we observed laid the basis for much of the Divergent Collaboration work that followed.

The first Air Force Divergent Collaboration project was essentially a workshop version of the original event that had started this whole creation effort. The challenge was "Information Visualization" and the workshop was planned over a three-month period and held in January 2011. Other than one of the original S&Es from AFRL who became the problem owner for this project, none of the participants were the same. We wanted to see if we could stimulate the creativity that we had seen with the original diverse group in a three-day workshop on a more specific challenge and develop solutions or opportunities that the Air Force could test or implement.

The preparation work was significant. Since we did not have a volunteer group like we did with the first meeting, we had to determine who to invite to the workshop that might give us the collaborative innovation that we had seen in the meeting. To do that, we took the problem through a significant functional decomposition process that involved the problem owner, the WBI IDEA Lab problem analysis group and the WBI facilitators that would help with the workshop. The functional analysis was used to determine which functions are critical for information visualization in Air Force related combat operations and operator decision making. Once the functions were determined, we had several sessions to determine which professions and/or expertise categories were skilled in those functional areas. For information visualization, functions like graphic design, marketing, language translation, forensic analysis, intelligence gathering and delivery and control system display were identified.

The next step was to identify professionals in these areas who we could invite to the workshop. That was fairly easy but getting them to commit to a three-day workshop in Dayton, Ohio in January felt daunting. It took us about a month to get our target of 15 invitees to commit to the workshop. We sweetened the deal by offering a $1000 stipend and all travel, housing and per diem expenses to the participants. Half took the stipend, almost everyone took the expense money and only two from large companies wanted no compensation. Because of federal rules, any government employee could not take a stipend. We were surprised by the enthusiasm of the people who accepted to be involved in the workshop. We really couldn't guarantee them that the experience would be interesting but we did emphasize the nature of the workshop and the important and military related nature of the challenge. Some were intrigued by the uniqueness of the event and others felt that contributing to a military related capability was important. It turned out that getting participants to commit was not that difficult after all and that was a major factor in what became a five year series of over 25 Divergent Collaboration projects. In fact, we reduced the stipend to $500 and it made no difference in getting participants to commit. The most challenging aspect was the participant's availability on the dates of the workshop which we had to set at the start of the project.

The stage was set in December 2010 and detailed planning for the first workshop took all of that month. A draft agenda was created with half-hour modules, break times, social events, break-out rooms, multiple team groupings, report-out processes, facilitation guides, information capture processes, problem user briefings to the group and many workshop facilitation ideas that came from facilitator guides and some that we had to design ourselves. While the participant and facilitator agendas were very specific, the following themes guided that workshop and all the Divergent Collaboration workshops to come. The first half day was introductory for the participants, the second half-day focused on the problem to be solved, the first evening was a planned social, the second day was a series of team break-outs to understand, analyze and dissect the problem, the second evening was a planned dinner with homework, the third day focused on synergizing the solutions, presenting solution opportunities and celebrating.

Over the years, we modified the agenda somewhat but the basic approach held. The overall concept was to get the diverse group of participants to really get to know each other, each of their perspectives and each of their talents. The problem was not exposed to the participants until the afternoon of the first day and we kept all of the participants in the problem space through most of the second day. Only the end of the second day and the third day were used to create solution opportunities (preferably broad opportunities rather than specific solutions) which the problem owner could then investigate after the workshop. We used various introduction approaches including story-telling, games and speed-dating to help with that. By the lunch break on the first day, the participant were very excited to get to know more about everyone and the buzz was palpable. Several described it as being invited to a very well planned dinner with guests that had extremely interesting backgrounds and getting to share their background and stories with each other. That was exactly what we were after. Several years later, we jokingly called ourselves great event planners but never actually offered that service.

The innovation aspects of the workshop followed many of the practices and processes that we had developed or used in more traditional innovation workshops. The unique and actually interesting aspect of this workshop was that none of the participants were experts or practitioners in this specific problem area and none had ever worked with each other before. The diverse and divergent approach really led to some fascinating solution opportunities, described later. The high-intensity pace of the workshop also contributed to a high-energy level of activity and the participants were eager to work to all hours and in any way possible to deliver possible solution opportunities. And the underlying collaboration characteristic of the event led to some continued relationships and partnerships after the workshop was over. While the problem owner, as well as future Divergent Collaboration problem owners, was very satisfied with the results of the workshop, the real surprise came from the participants. Everyone in this workshop commented that it was the best workshop that they had ever attended and that, besides coming up with useful solutions to a challenging military problem, they had a wonderful time and connected to some people that they would never have met. After twenty-five Divergent Collaboration workshops, we get the same feedback from almost all of the participants.

Over the five year period, we conducted some very interesting Divergent Collaboration workshops. Below is a short description of some of these workshops, the problem that was considered, some of the participants that were invited and a summary of the results.

Autonomy Divergent Collaboration, Aug 2012

Problem: Enhance the effectiveness of autonomous unmanned air vehicles

Participants: robotic surgeon, investment manager, peace corps returnee, astronaut, heavy construction manager, human-system interface designer, army warfighter, commercial airline pilot, drone hobbyist, instructional design teacher, cognitive systems researcher, entrepreneur, generational expert

Solution Opportunities: Conduct research on how different generations accept and utilize technology changes and greater automation in human systems, Consider cafeteria-style use structure and pick-n'-choose what levels of automation is needed or wanted, Implement greater automation in places that enable humans to do more than they could without it or where it's not possible for humans to control certain aspects of automation.

Quantified Warrior Divergent Collaboration, Sep 2013

Problem: Measure the human performance characteristics and potential of military warfighters

Participants: NFL trainer, police officer, consumer neuroscience expert, user experience expert, veterinarian, medical diagnostician, teacher of autistic children, personal trainer, Olympic judo coach, body language researcher, drug enforcement detective, high school teacher, pediatric physical therapist, emergency room flight nurse.

Solution Opportunities: Conduct failure profiles on how to understand behavior signatures just before performance deteriorates, Develop monitors to tell operators their state and how to compensate for deficiencies and whether or not to proceed, Examine team performance as a whole similar to how cattle ranchers look at the health of the herd to determine individual cow health.

<u>Energy Transfer Divergent Collaboration, Jan 2014</u>

<u>Problem</u>: Determine effective ways to heat fried food in an order of magnitude quicker.

<u>Participants</u>: anthropologist, thermal dynamics expert, anesthesiologist, spirits distiller, glass blower, firefighter, fluid dynamics expert, sterilization equipment designer, renewable energy researcher.

<u>Solution Opportunities</u>: Over one hundred solution opportunities were generated in this commercial-systems focused workshop including: exploring the use of additional microwave technology in the frying process, applying sterilization heating concepts, changing food separating procedures, use of laser and directed energy concepts for additional heating and restructuring the overall food preparation and management process.

In 2015, we invited a Harvard Business Review editor to attend one of our workshops as an observer. We wanted to get his opinion about the uniqueness and efficacy of our Divergent Collaboration process, but he soon asked to become a participant in order to get the full experience of the process. On leaving, he repeated what so many other participants have said, that it was a great experience and he had a great time while solving an important national problem. Several months later, one edition of HBR talked about our Divergent Collaboration process as a pioneering innovation approach that holds great promise for developing unique solutions to challenging problems. In addition to the HBR review, the WBI Divergent Collaboration process has been featured in several innovation journals, briefed at three national innovation conferences and the subject of two videos highlighting this unique collaborative innovation processes. While many IDEA Lab and AFRL personnel were involved in these workshops, Candace Dalmagne-Rouge was a key player and creator in most of the workshops and she documented, filmed and presented a number of important papers on the Divergent Collaboration process. What started as an awkward meeting between artists and engineers ended up as one of WBI's high-demand innovation offerings.

▎GRAPHIC REPRESENTATION

The powerful facilitation tool of Graphic Representation came to WBI's attention in a very unusual way. In 2015, the IDEA Lab hired a graphic illustrator to augment our workshop capture and reporting efforts. Jennie Hempstead was not only a graphic illustrator but a very talented artist with a degree in fine arts. We asked her to attend some of the workshops that we were holding for AFRL and to capture both the report-outs and the recommendations that came from the cluster groups in the workshop. As she captured the information on her notepad or laptop, we noticed that she was using unique icons, caricatures and other creative symbols as a holistic and short-hand method of representing the information. Since she was an artist, these drawings and graphics were outstanding, so we asked her to use the white boards in the workshop rooms and capture the information and reports in real time using her graphic representations for all the participants to view. From the outset, Jennie's illustrations were a great hit with the participants and the facilitators because it provided an additional representation of the report-outs and word charts that were being used by the participants. The more she used this skill, the more useful were the graphics and she was able to capture complex thoughts in real time on the white boards. As participants introduced themselves, she drew icons and pictures of many important elements of their expertise and experience on one large montage that stayed visible to all participants throughout the workshop. This allowed anyone in the workshop to quickly identify the talents and potential contributions of every participant at a glance and greatly enhanced the discussions in the workshops. We use graphic representation as often as possible in our workshops and now consider it an important tool and necessary technique for successful innovation workshops. Interestingly, graphic representation and graphic facilitation has now become a recognized process for meeting support and information capture and training and development programs are now being offered throughout the world.

THE SOLUTION SPACE

Einstein tells us that if he had 60 minutes to solve a problem, he would spend 55 minutes in the problem space and 5 minutes in the solution space. And we agree. But the Solution Space is pretty big too! So we spend a lot of time in the Problem Space but we cover a lot of territory in the Solution Space.

Once you have exhausted yourself (literally) trying to really understand all the aspects of the problem and you've arrived at a crisp, clear and concise problem statement, then jump into the universe of possible solutions. There are possibilities, opportunities, insights, ideas and actual solutions out there that will help with your challenge. A further examination of all the literature and research relevant to the challenge area sometimes led us directly to a solution opportunity. The global landscaping that we used in the problem space gave us many clues for where to look. If we couldn't find what we needed, we tried asking the world for help using crowdsourcing challenges. Often more research was necessary and we engaged the AFRL research teams to look for answers and insights. Once a possible solution concept was found, a simple prototype of that idea could be constructed, tested, iterated and refined. Starting simple, testing quickly, iterating rapidly and repeating the process is what rapid prototyping is all about and it proved very valuable for arriving at a viable solution. Again, collaboration in the Solution Space is just as important as in the Problem Space and there's a big world out there to look for partners.

CROWDSOURCING CHALLENGES

Crowdsourcing was a relatively new innovation technique when we started the IDEA Lab in 2005. Like most organizations that were in the innovation arena, we saw the benefits of crowdsourcing as an approach to problem solving and uncovering potential solutions. We utilized our own website for crowdsourcing and we contracted with several vendors to employ their systems and software for finding new ideas and solutions.

With a roster of 10,000 S&Es, we started by crowdsourcing the AFRL workforce and soon expanded to other government groups. At first, we posted a challenge on our website or their websites and hoped for a response. The initial response was not good, very few AFRL employees saw the value of responding and felt that this would take away time from their current work. We didn't do much better with other government agencies. Eventually, we offered prizes or incentives to solution providers but again the response was not very great. While we continued crowdsourcing in AFRL and we targeted similar organizations in the Navy, Army and NASA, the return on the investment was minimal. We concluded that we might be targeting the wrong crowd and switched to a broader and more global approach.

After working with several crowdsourcing providers, we employed InnoCentive for our crowdsourcing needs. InnoCentive is an open innovation and crowdsourcing system that enabled us to put Air Force challenges out to the world to address and sometimes solve. Since InnoCentive had a network of over a million problem solvers and offered monetary awards from $5,000 to $100,000 to the solvers, our challenges did get a lot of attention and some brilliant solutions from the most unexpected places were submitted. In some cases, we had to disguise the challenge to make it look generic versus military and several rounds of iteration on any given challenge often occurred. But the results were surprising and very valuable. AFRL covered the prize and crowdsourcing services funding and IDEA Lab personnel worked with InnoCentive on the posting, response management, selection and award determination elements. WBI posted dozens of AFRL problems with InnoCentive and found many useful solutions and identified quite a few unique problem solvers in so doing. If a solution was selected and rewarded, the offeror was then connected to the AFRL problem owner for further development possibilities. Several of the winners did receive substantial Air Force funding for prototyping and some actually found operation use.

One particularly successful crowdsourcing project led by Bob Lee was the jet fuel recovery challenge posted for AFRL by WBI through InnoCentive in the summer of 2012. The specific problem description that was posted was to enable recovery of jet engine fuel spilled from a storage silo. It was generated by a real Air Force problem that occurred routinely on Air Force bases and was very problematic particularly when the bases were remote from the continental United States. The posted challenge included specific requirements, such as, the solution system must remove synthetic surfactants that are commonly used in fire-fighting foams as well as water and other debris and that the solution needed to be portable as part of an overall fuel recovery solution that can recover up to 200,000 gallons in an 8 hour period and be immediately ready for use in operational aircraft. The posting had a $20,000 award for a practical and feasible solution and generated over 447 serious inquiries from 57 countries. After two weeks, 56 specific solution proposal from 13 countries were submitted and two $20,000 prizes were awarded to two individuals, interestingly both from the USA. One proposal was for a dual-stage filter system and the other was for a centrifugal separator. The proposals and submitted designs were so impressive that the Air Force Environmental Engineering program office built prototypes of both systems in order to reduce program risks. The prototypes were tested with Air Force JP8 jet fuel that was set on fire and extinguished with Air Force fire suppressant foam. Ninety-two percent of the fuel was recovered in the prototype tests and the recovered fuel successfully passed all Air Force and independent fuel industry testing for direct re-use in Air Force aircraft. The solutions were directly integrated into the Air Force's High Impact Prototype Program and delivered to the Pacific Air Force Command for operational use. While it is impossible to estimate the cost savings accrued from this crowdsourcing project, the total award of $40,000 is only a small fraction of what the Air Force would likely have spent to solve this problem. Overall, crowdsourcing and global innovation challenges have been a very successful solution generation technique and WBI continues to use InnoCentive for AFRL and its other customers.

RAPID PROTOTYPING

If there is one solution space concept that has proved itself to be extremely beneficial and valuable for helping WBI customers, it is rapid prototyping. It started as one way to build and test a potential solution that came from a workshop. Towards the end of a workshop, the participants were encouraged to build a quick version of one or more of the solutions that they felt might work. Initially they constructed the concept or solution opportunity from common articles available in our facility, mostly from cardboard paper, staples, string and paper clips. As the practice became more successful, we stocked up on Lego kits, hardware, small motors and information technology devices that allowed the construction of mobile or dynamic rapid prototypes. In 2006, we offered a modest amount of prototyping funding to each workshop group and the participants were able to shop for whatever they needed to build and test prototypes by the end of the workshop. As time went on, prototyping was an integral part of any innovation project at WBI and the need for a first-class rapid prototyping facility was apparent to Dr. Alok Das and the leadership at AFRL.

In 2007, we searched for a good place to build a rapid prototyping facility and found an old food warehouse on the north side of Dayton that looked interesting. It was located in an industrial park which housed machine shops and other small manufacturing businesses which we believed might be helpful in the future. Although the 25,000 square foot building that we chose left a lot to be desired, one of our best engineers, Mike Osgood, took it on and converted the warehouse to a first class rapid prototyping capability. With support from the owner and the city, the building was refurbished in record time and minimum investment. In 2008, we opened the facility and named it Tec^Edge Works. It was a perfect name since this facility, like the Tec^Edge facility nearer the Wright-Patterson Air Force Base, allowed us to build prototypes with leading-edge materials, advanced designs and unique features. The Works title was a reference to the Lockheed Skunk Works facility in Palmdale, California that had produced the SR-71 Blackbird, the B-2 stealth bomber and the F-117 stealth fighter for the Air Force. The actual operational aircraft were very large and complex systems, but the Lockheed Skunk Works had been used to create the original prototypes of these aircraft. In addition, the Skunk Works concept was to produce the prototype as fast as possible so the rapid prototyping culture is imbedded in any skunk works facility. The WBI Tec^Edge Works has operated non-stop since 2008 and has produced incredible prototypes for the Air Force, many of which have been tested in the field. A majority of the prototypes were versions of Unmanned Air Vehicles (UAVs) with unique sensor and navigation systems and quickly received endorsement from Air Force warfighters fighting in the Middle East. It was not unusual for these prototype systems to see operational experience in the field and several of the UAVs went immediately to production and deployment in the battle field.

While the Tec^Edge Works Rapid Prototyping facility or environment was often used to deal with real operational challenges and needs, its role as a part of the larger WBI innovation process is best illustrated by an example. As a response to stopping potential adversarial attacks on an Air Force Base, rapid prototyping at Tec^Edge Works was used to obtain solutions to stop an Air Force base penetration by an unknown automobile or truck while not destroying the perpetrators or their vehicles. Crowdsourcing had resulted in a unique design of a very flat ground vehicle that could be launched at an oncoming truck or car and then inflate itself to raise the intruding vehicle high enough so that's its wheels did not touch ground. This design stopped the oncoming vehicle from any forward progress and allowed Base guards to approach and see if the intrusion was accidental or intentional without harming the vehicle or its occupants. Rapid prototyping was then used to develop the sensors and navigations systems required for such a unique interceptor.

COLLABORATION ACCELERATORS

Another unique approach that was added to the IDEA Lab's innovation portfolio in 2014 was the concept of Collaboration Accelerators. Whether Collaboration Accelerators are collaborative innovation processes or innovative collaboration practices is debatable. Regardless, it has been a very exciting and creative activity that started in a tavern conversation with several members of the Dayton academic and collaboration community. After that conversation, planning was initiated to hold a special, three-month, summer program at the University of Dayton's Art Street complex that would involve twenty full-time students. The students would be selected based on their diverse interests and their willingness to engage in a creative exercise that was being developed as it was being implemented. Housing in the complex was provided by the University of Dayton and funding for the program was provided by AFRL. The students would receive training and facilitation by professionals from the Institute of Advanced Collaborative Training at the university and Wright Brothers Institute. The three month program was designed to create the conditions and environment for intense collaborative innovation by the students on real challenges offered from the Air Force and the Dayton Community. We called the program the Collaboration Accelerator because of its 24/7 nature and the full-time facilitation and learning support provided to the students. The background of the students ranged from theatre to engineering, arts to business and music to medicine. While most of the collaborative innovation workshops, like Divergent Collaboration, that we had organized at WBI were less than a week long, this program would last three months, operate 24/7, and would engage young students still broadening their skills as opposed to seasoned professionals and practitioners.

The first week of the program was primarily engagements with each other and training on the fundamental elements of collaboration and innovation. Challenges were then offered to the students and they were encourage to investigate these challenges, form small teams to focus on one or more challenges and then select two preferred challenges by the fourth week. The remainder of the ten week program was spent developing solutions to the challenge problems and presenting their solutions to the challenge owners and Collaborative Accelerator staff during the last week.

The first Collaboration Accelerator program was a great success. The two challenges chosen by the students were interesting: 1) How to warn runners or bikers about a vehicle approaching them from behind and 2) How to provide information about the Dayton Community to young newcomers in the most powerful and effective manner. The first challenge was provided by the Air Force but clearly had implications to the community which had just implemented a rent-a-bike program for tourists. The second challenge was important to the Air Force as well as to the Dayton region since AFRL was having difficulty recruiting new hires from the larger cities on the east and west coasts because of the image of Dayton as a sleepy, mid-west town. The students totally engaged on the challenges, came up with unique solutions and presented them in theatrical fashion at the end of the program. The "rear-view" problem was solved with novel head or bike-mounted sensors and warning systems provided by Air Force S&Es and the community challenge solution was presented in a one-hour play that followed a young newcomer's arrival in Dayton. Both solution sets were outstanding and imaginative. More important, the twenty students were extremely connected by the end of the summer, two students even fell in love and got engaged. The first Collaboration Accelerator had clearly proven that intense collaboration among a diverse set of students could generate extremely innovative solutions to tough problems. While full-time collaboration for ten-weeks is difficult to achieve in industry and government, the learnings from the Collaboration Accelerator can be applied to less intense innovation projects and the university and WBI are doing just that. The professionals from the University of Dayton, WBI and Dayton community that had created and supported the program were extremely pleased with the outcome and now sponsor a Collaboration Accelerator every summer

COLLABORATIVE INNOVATION

"No problem can be solved from the same level of consciousness that created it", Albert Einstein

| COLLABORATION AND INNOVATION

The Wright Brothers Institute has always been an Innovation Institute, but as we powered innovation, it became clear to us that Collaboration was the mother of Innovation. In 2009, we opened a new building with 23,000 square feet of space and called it the Wright Brothers Institute Innovation and Collaboration Center. All of the space was designed and used to foster collaboration. While innovation was always an objective of the teams and groups that met at the WBI center, we created an environment of collaboration where innovation could grow.

Every meeting room in the building used techniques and tools to foster collaboration. White boards, reconfigurable furniture, SOTA WiFi and high-end projection systems were standard. A large café offering food and drink with many spaces and tables to cluster and engage was built into the facility. A fully-manned operations center to satisfy any business need of the customer was available every day from opening to close. In essence, we built a conference center specifically devoted to creating and enhancing collaboration in order to meet the innovation needs of our customers. Between 2009 and 2019, over 150,000 people met in the Innovation and Collaboration Center to collaborate and innovate.

While the environment was very helpful, facilitation was just as important to energize innovation. Because collaboration and innovation are so intertwined, most teams needed help with both. With all good intentions, most groups assume that collaboration will occur when they set up a meeting and the main focus will be to produce the desired innovation. We found that collaboration can be very fragile and facilitators were often needed to create and maintain the collaborative environment. In a collaborative environment, all the tools, techniques and processes that facilitators can bring to produce innovation become more effective. The combination of environment and facilitation proved very successful for achieving collaborative innovation and those two capabilities became the foundation for everything we did at WBI. A collaborative environment and effective facilitation are powerful tools for building Collaborative Innovation and much of the work of the Wright Brothers Institute has centered on using those tools to help our customers.

| COLLABORATIVE INNOVATION TOOLS

There are many techniques cited in the literature that are very helpful for creating collaboration and innovation and we have used many of them in facilitating our meetings and workshops. But three rather unique tools have been particularly effective and have really delivered unexpected results: Talking Sticks, Behavioral Types and Curiosity Development.

Talking Sticks have been around for millennia. Cave paintings of primitive tribes show articles being passed from one person to another in a group setting. While we can't be sure that the depicted gatherings were promoting team building and collaborative decision making, more modern tribes certainly used that technique to achieve agreement on important matters. And the object that was passed wasn't always a stick. But again, many tribal cultures from the last two millennia did pass jeweled or elaborately carved pieces of wood to each other when they were trying to arrive at a critical decision. And so we settled on a beautiful Talking Stick made expressly for us by a Shaman/Jeweler in Santa Fe, New Mexico. Ross LewAllen had spent six months of every year from 1970 to 1990 visiting and learning the customs of primitive tribes from all over the world. In every visit, he observed the use of a Talking Stick by the tribe when important decisions had to be made. So when he made a stick for us, we figured it was pretty authentic.

When using a Talking Stick, there are only two rules that people have to follow. Whoever holds the Talking Stick must "Speak from the Soul", that is, speak the truth, the whole truth and nothing but the truth. Everyone else must "Hear from Heart", that is, really listen to the speaker and be appreciative and understanding of what he or she is saying. After the holder has said his piece, the Talking Stick is passed to the next person who must follow the same rules. These rules must be agreed to by all parties before the talking begins and there really isn't any way to check to see if anyone is cheating. It's a promise and commitment and generally everyone observes the rules. In jest, we sometimes told the Talking Stick participants that if they didn't follow the rules, the Talking Stick would get hotter and might burn them.

We used the Talking Stick when we really needed to get to issues that weren't be addressed or were purposely being avoided. This behavior can destroy a team and true collaboration cannot exist in that environment. We also used the talking stick to accelerate the pace of the meeting since it stimulates the participants to get down to business by truth telling and active listening. While just a simple tool, participants generally agree to abide by the rules and get very serious when it is used. We often started meetings with the Talking Stick because it set the nature of the meeting to one of seriousness and importance. The Talking Stick has been a powerful tool to enhance both collaboration and innovation in team meetings and workshops and its unique nature makes it a popular addition to all the other team building tools that we employ.

In any gathering, much of the specific information about a participant is shared at the outset of the event. Often, some of this information is already known to some of the participants but if not, there are many techniques to introduce participants to each other. Besides the standard information, we have used ice-breakers to extract unique information about each participant. Regardless of how much time is devoted to this, it still doesn't get to some of the most important knowledge about a person. We know this because at the end of a workshop, participants tell us that they would have really enjoyed getting to know some participants better but didn't know how to approach and connect with them. Because this is so prevalent in business meetings that are focused on innovation, we have routinely used behavioral type instruments, such as the Myers-Briggs Behavioral Type Instrument (MBTI) to help participants with communication and it has also helped with achieving and maintaining collaboration.

While some people see it as categorizing people, we have used behavioral type instruments to help participants understand and communicate better with each other. MBTI, in particular is so popular that participants often know their type (e.g. ISTJ, ENFP, etc.) Regardless, we often ask participants to voluntarily take a very quick version of the instrument which gives them their type in 15 minutes. And of course, our facilitators reveal their type to the group. In anyone in the group does not want to take the instrument or does not want to reveal their type, our facilitators are trained to observe and identify the most common types in any person. For example, if they need time to think before they talk, they're probably an "I" or if they talk as they think, they're likely an "E"; if they are near-term focused, they're often an "S" and if they are future-focused they're likely an "N". We found that most of our clients from AFRL and the S&E world make decision by thinking (Ts), not emotionally (Fs) and they like rules and structure (Js) versus being very comfortable with change (Ps). Generally, our workshop participants are ISTJs, INTJs, ESTJs and ENTJs. Quiet versus Talkative, Tactical versus Strategic. Even if we don't know their specific behavioral type, just understanding that many of the participants have these preferences is very helpful. For collaboration, knowing how people generally receive information, process that information and make decisions using that information can be very helpful. The same applies to innovation. It is a powerful technique that can save great amounts of time, get to the meeting objectives more efficiently and enhance the engagement and participation of all participants in the meeting.

A third tool that has proven very useful is Curiosity Development. There are some people that are just naturally curious and will immediately begin asking questions, googling anything that they don't understand and turning over any stone that gets in their way. These are the best participants for a Collaborative Innovation workshop because they will naturally be interactive, participative and creative. The real challenge with curious people is to keep them focused on the primary objective, but facilitators are trained to handle that.

A more difficult challenge is developing those participants that are either not curious or only mildly inquisitive. We use a number of techniques to accomplish that. One has already been mentioned, we keep them in the problem space and tear the problem apart. While some participants get upset or anxious, it's hard to stay in the problem space without asking questions. Once a participant gets engaged, that curiosity builds and the participant becomes a model for others to engage.

Another is to purposely add participants to the workshop that have very unique perspectives, expertise and experience. This not only helps with innovation, it also contributes to collaboration. Most people are interested in learning new information, new skills or new opportunities. That's probably why news programs dominate television and radio. If all of the people in the room know everything about each other or have the same perspective on the subject, there's really no need to ask questions or probe for information. By diversifying the group, curiosity becomes a useful way to participate and learn.

A powerful way to get participants engaged and pique their curiosity is to use storytelling whenever possible. Communication between people is often reciprocating information flow. I tell you something and then you tell me something else. If this is all that happens in a meeting, there will be very little collaboration. But if participants tell stories to each other, about themselves, their work, their best experiences, their worst failures, it begs for questions. However people tell stories, it's always from their perspective and not the perspective of the listener. So the listener needs to be curious and ask questions in order to understand the story. Again, this has to be facilitated because people get so excited about telling their story and listeners get so interested in some of the details that they could unintentionally derail the meeting. In many of our workshops, we have to forcefully interrupt or end the story telling modules because people are so engaged and would go on forever. Storytelling has many positive attributes and almost invariably develops the curiosity of all the participants.

COLLIDERS

In 2014, WBI became increasingly involved in supporting and enhancing the Air Force's Small Business Innovation Research (SBIR) Program. We had conducted a twelve month study to determine the reasons why the Air Force SBIR program was not meeting expectations and found that one principle issue was communication between the Air Force and small businesses as well as small businesses with each other. We set up a Small Business office inside WBI and chose a Director to manage it. The first director was Candace Dalmagne-Rouge who had been a principal in the development and execution of the IDEA Lab's Divergent Collaboration program. After a year, she moved on to an outside start-up development organization and Jim Masonbrink took over the WBI Small Business office. Both Candace and he dealt head-on with the issues that we had found from the study and created a program to connect the various stakeholders in the SBIR program. The program focused on a well-organized series of collaboration events called Colliders.

Colliders are essentially open-invitation events that focus on some aspect of the SBIR program. They are well advertised in advanced, open to anyone interested and organized to provide important information on one aspect of the SBIR program. For example, a topic might be on the process that the Air Force uses to solicit proposals on a technology challenge that they would like small businesses to investigate. The Air Force managers responsible for that area would be invited to start the Collider by briefing the attendees on the process and any information that might help a small business in developing their proposal. While that aspect of the Collider is important and stimulates attendance, a major objective of the Collider is to facilitate the interaction between all attendees after the presentation. Space is set apart for an hour or so after the presentation for attendees to interact. Sometimes, food and drink are included and there are tables and chairs for the attendees to cluster and exchange information. While it seems an obvious step to encouraging collaboration and even innovation between anyone interested in Small Business Innovative Research, this was the first time that an organized Collider program had been implemented. It was an immediate success and had unforeseen benefits for both the Air Force and the small business community. Better proposals were an immediate outcome since the small business had a better idea of what the Air Force wanted and how they could submit their ideas. Collaboration between two or more small businesses that had been initiated at the Collider led to more multi-disciplinary technology proposals. As the small business attendees interacted with the Air Force SBIR managers at the Collider, new ideas for future solicitations by the Air Force were generated, which led to greater responses by the small business community. The Colliders were a win-win for everyone attending and they were in great demand by the small business community. The WBI Small Business office developed a schedule for monthly Colliders and leveraged the internet to broadcast it far and wide. Eventually, the Colliders got so popular and were so large and frequent that WBI and the Air Force partnered to create an additional collaboration center in downtown Dayton to house the Small Business office and hold the Collider program. Colliders are a great example of how collaboration and innovation can be stimulated and supported by simply providing the right environment and a managing and facilitating the process to focus on those outcomes.

| COLLABORATIVE OUTREACH

The success of the Collider events and the need to support that activity in a larger collaborative environment led to the creation of the WBI Collaboration Center in downtown Dayton. In 2016, AFRL provided funding to WBI to create a collaboration center in the heart of the city of Dayton that could be more accessible to both small and large businesses that were interested in working with the Air Force and AFRL. Jim Masonbrink transformed an old warehouse into a unique environment where Colliders and other collaboration events could be held. The Downtown WBI facility at 444 Second Street had all the operational features and amenities to encourage and support collaboration and innovation. Besides providing space for networking and small rooms for collaboration, the environment had an edge to it that appealed particularly to entrepreneurs and young people. The WBI space was adjacent to that of the Dayton Technology Entrepreneur Center (TEC) and often the activities of both were combined. The number of Colliders significantly increased in this facility and the WBI Small Business office was moved to 444 Second Street. In addition, WBI and TEC programs and projects were often held at this facility and the interactivity between the two definitely supported the collaborative outreach of both WBI and AFRL to the Dayton community. Despite the fact that the airplane had been invented in Dayton, Ohio by the Wright Brothers and Wright-Patterson Air Force Base is the largest Air Force Base in the world, there had never been a physical Air Force presence in downtown Dayton. This outreach by AFRL through WBI created a physical portal for anyone to connect with WBI, AFRL and the Air Force in a facilitated way.

From 2016 to 2020, hundreds of events, programs and workshops were held at WBI's Downtown facility. The outreach of the Air Force to the city of Dayton, the Dayton Regional community, the State of Ohio and the entire country through this facility created collaborations between the Air Force and the civilian community that led to important collaborations and significant innovations. Improvements and additions to the facility were made over that three year period to accommodate increased usage and novel approaches to collaboration and innovation that were stimulated by the environment in this facility. In 2019, the Air Force increased its support to WBI and expanded the WBI presence at 444 Second Street to accommodate a significant part of the Air Force SBIR office, the AFRL Technology Transfer office and the WBI Commercialization Services office. The facility was completely redone to accommodate the latest innovations in outreach and storefront best practices and re-opened in 2020. The pioneering work of the WBI collaboration and commercialization services groups coupled to the determination of the AFRL Small Business Office to reach out and network with the outside community has clearly been a breakthrough in collaboration innovation.

| MAKER SPACES

While it's not exactly clear how and why this happened, the S&E workforce of AFRL slowly lost its own ability to rapidly make and test simple versions or prototypes of potentially innovative systems to satisfy Air Force needs. As AFRL received greater amounts of funding from the Air Force budget, it was easier to contract out to industry and academia for technology development and less and less R&D was being done internally by the AFRL workforce. The larger budget also made it easier to hire support contractors to do some of the work that had been done by the organic workforce. While the budget increased the manpower level was fixed and the percentage of AFRL personnel doing R&D kept decreasing. Finally, in-house laboratory facilities were difficult to maintain and support with technicians and their number and availability also decreased. By 2010, the amount of in-house R&D was less than 25% of what it had been in 1980 and neither the facilities, personnel nor incentives for conducting hands-on research and development were prevalent in AFRL. The result was a workforce that could manage R&D but did very little of it themselves and this was not what the Air Force needed.

As a start to changing this trend, in 2016 WBI and AFRL built a modest Maker Space in one of the rooms at the WBI Innovation and Collaboration Center near W-PAFB. This Maker Space was designed by Dr. Emily Fehrman Cory of AFRL and later managed by Tom Mitchell and Joe Althaus of WBI. A Maker Space is exactly what it says, it's a space to make things. Many universities now have Maker Spaces on their campus and communities are now building Maker Spaces for the general public to use. In Dayton, there is a Proto Build Bar which combines a maker space with a meeting space that includes a bar. The WBI Maker Space was outfitted with several 3D printers, shop machines and tools, small motors and Raspberry-Pi computers and plenty of bench and work spaces for constructing rapid prototypes of small devices and products for testing and experimentation. More importantly, we staffed the Maker Hub with individuals that could help anyone use the equipment and assemble the devices since this could be intimidating to a first time user.

At first, only a few AFRL S&Es trickled in but the utilization of the Maker Hub increased with time. Once an S&E became comfortable with using the space, they came back and often stayed to the point where we had to ask them to leave at the end of the day. The use of the Maker Hub increased to the point where we had to double the size of the Maker Hub in 2019, increase the staff to two full-time master makers and add very high end Laser Printing Systems, complex computational subsystems and "tool and die shop" level machinery. Besides increasing the competency, curiosity and capabilities of the AFRL workforce, the WBI Maker Hub is heavily used for WBI Tech Sprints since it can produce modest prototypes of potential solutions in hours to days.

| TECHNOLOGY SPRINTS

Technology Sprints have been part of the WBI innovation process since the beginning of the IDEA Lab. Workshops to develop solutions to solve operational Air Force problems often led to preliminary designs that could be constructed very quickly and tested before the workshop was concluded. But the concept of starting with the promise that a preliminary design(s) or rapid prototype(s) would be built before the workshop concluded was somewhat unique. Google had experimented with Sprints to rapidly prototype some of the ideas coming out of Google-X and had honed the process into a practice. We came up to speed quickly on their process and found that it closely resembled our more aggressive workshops that resulted in a rapidly constructed prototype. At WBI, Tech Sprints are generally led by Joe Althaus and Tom Mitchell and are nominally a week-long workshop with significant preparatory work and meetings. The problem owner has to be clear on what he/she wants and must buy into the concept of a Tech Sprint. The owner is asked to kick-off the Sprint on the first day, be available throughout the week if needed and present for the final reporting and demonstration of the prototype on the last day. Once the owner describes the problem, the usual techniques of problem decomposition, functional analysis, global landscaping and intelligence analysis takes place. The Sprint team is kept relatively small and includes the problem owner, several SMEs that are working or have worked in the problem area, one or two individuals that have experience in the Sprint process, at least one WBI facilitator and a WBI expert, usually from our Maker Hub, that can help to make a prototype.

By the middle of the week, the team has engaged with outside experts and has developed one or more solution opportunities for further exploration. The most exciting Sprints have identified at least two solution opportunities and the participants break into smaller teams to work on each of them. Most of the rest of the week is spent designing and building and testing the prototype and the last day is taken with presentations and demonstrations to the problem owner and any stakeholders in the problem that he/she has invited. Often, parties that may be interested in further developing or even commercializing a prototype solution are involved and provide a pathway to achieving a solution. While the preparation for a Sprint may take a few weeks, the week-long Sprint is an excellent way to develop a possible solution quickly and create a prototype for further exploration.

As an example, WBI partnered with the AFRL Human Performance Wing, a national healthcare provider and the City of Dayton to examine an important problem created by opioid addiction. The Dayton Region has been described as one of the worse opioid addiction regions in the nation and its leaders were very interested in working with WBI on any aspect of this problem. The Air Force has always been vigilant around opioid addiction because many of the wounded warfighters have potential opioid dependency because of the pain treatment and pain management that they have received on returning from the battlefield. This Sprint, which was stimulated by a nurse from a health provider, was to investigate innovative ways to prevent a person with an opioid dependency from using the Pic Line that was often placed on a hospitalized addict to administer medicines both in and after release from the hospital. While it seems surprising to a non-addict, the Pic Line is a convenient and simple way for the addict to introduce an opioid directly into their body. In fact, many addicts use this method while still in the hospital if friends bring them the opioids.

The nurse problem owner had several ideas on how to approach this problem and several hours of background information which he shared with the Sprint team on day one. The problem decomposition and functional analysis led to several potential pathways and a global literature search revealed ideas on this issue that were being implemented for other related problems. The team had several late night sessions and came up with two different possible solutions. One involved securing the standard Pic Lines from non-prescribed use using lock-down systems and the other involved changing the design and operation of the Pic Line to prevent misuse. Both led to simple prototypes which were presented to the problem owner, stakeholders from the healthcare provider and several small healthcare business investors. Both prototypes were well received and both approaches were pursued after the Sprint, with one receiving venture capital funding support. This is a good example of what can be achieved in a five day Sprint and is typical of what we have seen in conducting more than thirty Sprints over the last two years.

| COMPETITIVE INTEGRATED INTELLIGENCE

The success of Front End of Innovation is critically dependent on knowing what information is out there that could be used to create innovative solutions. While traditional competitive intelligence has been greatly enhanced by the advances in information systems and search engines, in 2018 the Wright Brothers Institute's Larisa Dmitrienko and Heidi Longaberger partnered with CHN Analytics to create ML-CI2 (Machine Learning-Competitive Integrated Intelligence). ML-CI2 integrates cutting-edge machine learning-based data analytics to deliver insights from extremely large data sets and deep professional expertise to quickly cut through the noise in big data and identify high value answers to the cross-disciplinary complex challenges faced by its customers. WBI used this capability to assist several groups in AFRL and the Department of Defense by delivering a comprehensive awareness of multi-dimensional technology spaces and a facile way to visualize and understand these large data sets.

By using ML-CI2, WBI has been able to offer comprehensive

technology landscape analysis, creating a picture of who is leading in a certain technology area in a period of weeks versus case studies that can take months. To make it user-friendly, the information is presented in four categories: State of the Art of the Technology, Emerging Science and Technology, Benchmarking Information and Collaboration Possibilities using Tableau. This has helped WBI's customers to choose the right partners, conferences and publications to enhance trust, collaboration and innovation. In addition, ML-CI2 allows rapid updating and refreshing of the data for further analysis by the customer. Most important, ML-CI2 is a tool that is profoundly valuable for creating the partnerships and opportunities that are the basis for Collaborative Innovation.

STOREFRONT ENVIRONMENTS

The concept of "storefronts" was investigated several years before the creation of WBI and AFRL's own storefront at 444 Second Street in Dayton. In March of 2017, WBI facilitated a workshop to discuss and create mechanisms to connect the Air Force acquisition and research communities to individuals and teams who might be able to provide innovation solutions to Air Force challenges. WBI had been involved in the creation of the SofWerks facility in Ibor City, Florida, close to the Special Operations Command (SOCOM) Headquarters in Tampa, Florida. After conducting a Divergent Collaboration workshop for SOCOM with our sister innovation institute, The Doolittle Institute (DI) in Fort Walton Beach, Florida, SOCOM and DI personnel asked WBI to help them design a walk-in storefront where anyone with an interest could interact with SOCOM operators, gain a better understanding of the operator's problems and eventually submit ideas, prototypes or designs which might solve the problems through the storefront. SofWerks was the first storefront utilized by a major military command and it was a huge success. The location was near a community college in Ibor City and the design and operation was very welcoming. Soon, it was attracting innovative people from all over the country who had ideas that they could discuss with the SOCOM operators, get direct feedback from the users and then submit proposals for funding from SOCOM to develop the ideas. SOCOM then did something extremely innovative and positioned an acquisition official in SofWerks to review a proposal on the spot, sign a contract with the offeror and provide funding immediately to the proposer. To this day, SofWerks is the gold standard for storefronts and are very proud that we contributed to its success. We also learned many lessons from this experience that we applied to our own storefront facility.

The WBI Downtown Dayton storefront began as a place for small businesses to learn more about the Air Force and an environment that would support innovation workshops for WBI. It succeeded on both counts with the WBI Small Business Office conducting many Colliders in the facility and unique projects like the Summers of Collaborative Innovation, Technology Sprints and WBI-facilitated innovation workshops involving many participants from the community rather than the Air Force. For two years, it was the WBI hub in downtown Dayton and a great place to support collaboration meetings, week-long innovation Sprints and diverse perspective innovation workshops. While it was a well-designed and operated storefront, it was only modestly successful in drawing people from outside the community or the Air Force at Wright-Patterson Air Force Base.

As mentioned, the decision to place a substantial part of the Air Force SBIR Office, the AFRL Technology Transfer Office and the WBI Commercialization Services office in the facility changed all that. WBI leased the entire first floor of the building and created offices, meeting spaces and new information technology support systems that have made it the go-to place in the Midwest for small businesses interested in doing business with the Air Force. With the inclusion of the SBIR office, the experience of potential solution providers will be much more like that of SofWerks with the planned interactions with Air Force problem owners and the possibility of actually getting a contract and funding on the spot. The addition of the Air Force Technology Transfer Office and the WBI Commercialization Services office will draw Air Force S&Es from the Air Force Base to WBI Downtown since they can get help on how to patent their ideas and commercialize their inventions. The new environment is even more conducive to collaborative innovation than before with high-end technology capabilities and additional operational support. Finally, with tremendous help from AFRL, WBI has created a first-class storefront for the Air Force that will serve all of the mid-western region of the country.

| SUMMERS OF COLLABORATIVE INNOVATION

The facility that WBI designed and created at 444 Second Street was, in the vernacular, Cool! It had high ceilings, painted floors, garage doors to separate different meeting spaces and a huge retractable curtain made out chain metal to allow our users to see what was going on in the rest of the facility. From the beginning, everyone visiting the place felt that it had the right feeling for an innovation hub but we weren't quite sure how to take advantage of its unique environment. In 2017, one of the leading scientists from the AFRL Aerospace Vehicle Division came up with a great idea. He had studied Google's Summer of Technology process and felt that he could hold a similar event in 444 that would bring brilliant scientists and engineers from all over the world to spend a "Summer of Innovation" in the facility to solve a critical software problem for Air Force UAVs, assuring that their software can be verified and validated as they adapt and respond to their operational environment their operational environment. To do this, he needed the environment provided by the 444 facility to create a collaborative workspace for the fifty S&Es and all the equipment and support that an organization like WBI could provide over a 10 week period in the summer of 2017. He also needed about $1 Million to cover the costs and expenses of the participants who would come from all over the world. In a three month planning period, WBI modified the facility to his needs and he got the $1 Million from the Executive Director of AFRL.

That Summer of Collaborative Innovation at the WBI Downtown Dayton facility was a great success. Thirteen teams created from the participants worked until all hours throughout the summer and collaborated to solve one of the toughest problems for UAV software. The 50 participants loved the 444 environment and all of the features that allowed them to be innovative, collaborative and connected. An output briefing to the leaders from AFRL resulted in additional funding being dedicated to this problem area and several national funding awards were received by some of the teams. But the most important result for WBI were the best practices and lessons learned from a construct like a Summer of Innovation. We learned how easy it was to create collaboration in this environment but also saw that a major problem was capturing, coordinating and communicating the vast amount of real-time information that is generated in such an environment. We were surprised that there was little noise and clamor from 50 people in a closed environment. These S&Es, many of whom were software engineers, were at their best when they were quiet. They quickly set up a Slack hub that they used to communicate and collaborate 24/7. While it may have been quiet, observers could feel the energy in the group and facilitation, while available, was only sparsely used. This first Summer of Innovation was a great success, so we opened up the facility to more experiences like that from then on.

The second summer program at 444 was labeled the "Summer of Topological Data Analysis (TDA). Like the first, this one was held in the summer to take advantage of the better availability of students and faculty to attend. Most of the Summer of TDA participants were students who had shown an inclination or interest in the field of Topological Data Analysis. This information analysis area is particularly innovative because it essentially looks at information maps in three dimensional space and discovers insights and trends that enhance standard information analysis. Because they were mostly students, the summer activity was more like a giant term project where the students were learning and simultaneously contributing to the state of the TDA art. Seven very diverse data fields, from finance, to training, to people characteristics were chosen for application of TDA by the students. The students loved the idea of solving real world problems and being trained in this very powerful field of data analysis by some of the creators and world experts in TDA. Every week, one of these experts would spend a half-day with the students, who again had formed into smaller teams in order to focus on some of the unique aspects of the data fields and TDA concepts. Like the previous year, the

Summer of TDA was a great success, with outstanding final reports, students changing their majors to focus on TDA and several on-the-spot employment offers by AFRL.

As we grow as an institute promoting Collaborative Innovation, the concept of a dedicated group of highly-talented S&Es collaborating on a tough problem for an entire summer will be used more and more. Obviously, dedicating an entire summer with 50 people working on a problem can only be used in certain circumstances, but it is a very powerful tool that bears consideration if you really want to collaboratively solve a tough challenge.

▍HUMAN PERFORMANCE COLLABORATIONS

In 2018, the 711th Human Performance Wing of the Air Force decided to try a new approach to strategic planning and development of advanced human performance and enhancements technologies. They asked WBI to help with this endeavor and a three month program to accomplish this task was initiated. The approach was to hold twelve one-day workshops that would focus on a specific Core Technology Area (CTA) of human performance technology, discuss the needed Air Force capabilities in that area, brainstorm potential R&D pathways to develop these technologies, identify partners for collaborative activity and produce a first-cut strategic development plan for each CTA. We held all twelve workshops in a three week period and invited everyone in the Human Performance Wing to attend. For most of the workshops, over 100 S&Es attended and all were eager to contribute.

With such a large group, a tight agenda and very interactive facilitation was required. Even the introduction of participants was managed and took over an hour. Significant background information was provided to the participants before the workshops and lunches were available to encourage participants to work over the lunch break. The facilitation challenge was to allow all participants to contribute as well as interact with each other. To do this, we tried a simple approach of providing the participants with 8x10 inch post-it notes and large markers to write their comments. After a one hour briefing on the technology area, military requirements and current R&D efforts underway, participants were asked to suggest ideas and projects that would help with the technology development in that Core Technology Area, write them on the post-it notes and place them on the walls of the meeting room. Over lunch, the participants were able to look at the notes and discuss them with the note owner. During lunch, the facilitators and the AFRL CTA leads identified participants for inclusion in twelve teams of six to eight for group discussions. During the afternoon, the groups met, developed further ideas and projects, posted them during three rounds of discussion, the participants changed groups if they wanted and the workshop concluded with an enormous amount of insights, suggestions, projects and pathways for the CTA. After each workshop concluded, the AFRL CTA lead used this information to create a strategic plan for that CTA and briefed to the Human Performance Wing leadership on that plan within a week. The entire process took less than a month and resulted in a very useful and informed plan for the development of that CTA over the next five to ten years. While this particular Collaborative Innovation process is not applicable for the planning of every R&D endeavor, it is relatively easy to implement and is very effective at creating a challenging strategy and plan in a very short period.

▌ CORPORATE COLLABORATIONS

Until 2016, almost all of the WBI Innovation and Collaboration support was created to assist the Air Force Research Laboratory with innovation and collaboration. The organization and implementation of the IDEA Lab, the development of unique innovation and collaboration processes and workshops and the interactions and focus of WBI were always focused on the AFRL customer. By 2016, however, the work of WBI in Collaborative Innovation was getting significant publicity in the region and we were being approached by several organizations for potential support and services. Our mission clearly was focused on the Air Force and AFRL and we never wanted to diminish any support for these organizations by servicing other customers. After much discussion with AFRL and our Board of Directors, we decided to help a few selected companies in the Dayton area with collaboration and innovation support on specific projects. While we were paid for these services, there were two objectives to this work: 1) would our various collaboration, innovation and collaborative innovation processes that we had developed for an Air Force R&D organization (AFRL) be useful and applicable to the commercial and business world and 2) could we improve our Collaborative Innovation support to the Air Force from ideas and insights that we would learn from these commercial ventures. During 2016 to 2018, we engaged in projects with the Emerson Corporation, Midmark and Henny Penny. The projects were quite successful, so we concluded that we could use our processes to help the commercial world. More importantly, there were some important insights that we gained from applying our processes to commercial companies. The Emerson program is a good example of that.

Emerson is a world-leader in developing and applying climate management systems for industry and home use. They are based a few miles from Dayton and in 2015, created an Emerson Innovation Center in the heart of Dayton. During the design and development of this Center, they were frequent visitors to WBI and used many of our environmental ideas in their Center. When they began the operation of the Center, they asked WBI to assist them with a couple of their challenges. An interesting one focused on creating a home climate system that delivered the perfect heating, cooling and humidity environment to every person in a home at all times. Obviously that would be a desirable situation to a home owner and they might pay a premium to have that in their home. The challenge was how to have different environments in the various spaces of the home that would satisfy the individual preferences of the person in that space. At its best, a person sitting on a couch needing a warmer climate would receive higher temperature, higher humidity air than a person who had just completed a workout sitting two feet away.

We proposed a series of workshops that involved global landscaping and problem analytics to determine the participants for the workshop. We used all the innovation processes that we had employed with AFRL challenges and even included a Divergent Collaboration workshop to consider unique inputs. The program lasted about three months with four workshops in the Emerson Innovation Center and generated many ideas, solution opportunities and possibilities for Emerson to consider for meeting that home climate approach. The Emerson folks were happy with our processes and support and we felt that we had verified that our processes could be very useful and impactful for a commercial client.

Satisfying the second objective was less clear. We did learn a lot about how commercial companies approach innovation and collaboration. While Emerson's goals were similar to our experiences with the Air Force, two unique aspects were enlightening. Collaboration in the Air Force was generally acceptable as long as the individuals saw value in the time and effort needed to develop a collaboration, that is, it was more of a personal decision to collaborate. In business, collaboration seems to be driven by a directed or self-evident need to collaborate, that is, it is more of an organizational decision that collaboration is needed. Secondly, innovation is driven by a relatively near-term need to produce something that has an advantage over a competitor's approach, that is, the market and competition drives the innovation process. With the Air Force Research Lab, innovation is a continuing process that is always on the minds of the S&E because the competition is somewhat unknown, that is, there is an ongoing need to always be the best and never get into a situation where a serious threat could not be countered. These are significant learnings and we found them to be true with the other commercial clients that we supported. While we have not continued our commercial work, we feel it was successful and prepared us as we began to assist the more near-term, product-oriented part of the Air Force, the Air Force Materiel Command's Life Cycle Management Centers.

▌ MILITARY COLLABORATIONS

Since most of the WBI Collaborative Innovation services from 2000 to 2018 were focused on helping the Air Force Research Laboratory, it seemed logical for us to expand our activity to support the Air

Force acquisition community, in particular, the Air Force Materials Command's Life Cycle Management Centers. Our work with the commercial clients had given us some insights into how these more product oriented organizations in the Air Force could use our services. As with the commercial clients, we began by asking organizations within these centers to tell us about their most critical needs or their biggest problems. They were quick to answer these questions but they weren't sure if we could help them. Their traditional approach was to depend on their prime contractors to respond to their needs and to rely on their own engineering staffs to identify the near or far-term problems that would compromise their systems or operations. They were responsive to problems and not generally pro-active in seeking out potential issues. They relied on their own staffs to identify and solve some of the problems and if that didn't work, they primarily looked to their long-time prime contractors (like Boeing, Lockheed-Martin and Northrop-Grumman) to identify problems and propose innovation programs. This approach had been followed for decades and it was clear some new thinking, ideas, partners or opportunities might prove valuable.

From 2017 to the present, we have engaged with system program offices and operational teams in the Air Force and Special Operations Command to develop collaborative and innovative solutions to both identified needs and problems as well as unanticipated issues and potential challenges. These organizations have very structured processes to deal with requirements and needed improvements but most of them involve the same government or industry personnel that have generated the requirements and needed improvements. What we offered was to insert new thinking and processes into that approach to see if we could help them arrive at more creative responses or solutions in less time or for less money.

Our approach was very similar to what he had used with AFRL organizations looking for near-term innovation and what we were using with our commercial customers. The process started with decomposing the problem or need using our IDEA Lab process, determining the functional aspects of the challenge, identifying companies and individuals who had expertise in that area, organizing a workshop to include the problem owners, the government and industry experts who were relevant to the problem and a significant number of non-traditional players who might advance or stimulate the participants to look for new solutions. The workshops were well planned and organized and the government problem owners were generally excited about the new possibilities. The biggest challenge came from the government and industry participants who either thought they had the solutions or were typically the ones who provided the solutions. The addition of individuals from small businesses that were unknown to them, participants who had expertise in peripheral or tangential R&D areas or individuals that approached similar problems from very different perspectives was new and somewhat uncomfortable to them. At first it delayed or circumvented the innovation potential of these workshops but eventually most saw the merits of the process.

What really made a difference was the coupling of Sprints to the workshops. If a workshop came up with one or more ideas or solution opportunities, making a minimal version or rapid prototype of the potential concept really opened up the eyes of the government and industry participants. While these prototypes rarely solved the entire problem, they provided interesting insights that allowed the government and contractor to make quick decisions about the next steps. Several of the Air Force and SOCOM problem owners were very happy with this approach because it not only saved them a lot of money but accelerated the development of a solution or idea. One of the government problem owners told WBI that it saved them millions of dollars and years of time getting to a needed capability.

| CAPTURING OPPORTUNITIES

In 2018, WBI was asked to facilitate a workshop for the AFRL Strategic Development Planning and Experimentation (SDPE) group. The premise for the workshop was very novel. The project was initiated with a "Hunch" articulate by the Commander of the Air Force Training and Education Command (AFTEC). He stated that he had a hunch that by using Artificial Intelligence (AI), Virtual Reality (VR) and Augmented

Reality (AR), the training of a new Air Force officer to fly an F-22 advanced fighter aircraft could be reduced from 6 years to 1 year. This was an incredible statement to most Air Force and industry experts and particularly pilots who had gone through those long years of training to fly high performance fighter aircraft. On the other hand, why not look into this possibility to see if some of the training time could be reduced. SDPE labeled this activity as "Capturing an Opportunity" and WBI saw it as a wonderful way to use our innovation processes to go after a BHAG (Big, Hairy, Audacious Goal).

Since the premise was that AI, VR and AR would really increase the opportunity to reduce the training time, we immediately investigated these areas for experts and experience that might bear on this problem. It wasn't hard to find many organizations doing R&D in this area. In fact, that was a minor problem because many product development companies assert that they incorporate AI, VR, and AR in their offerings when in reality it is just advanced information technology. We focused on the education and training industry since this was the subject of the Capturing Opportunities project. Even in this subgroup, the application of AI and AR ranged from cosmetic to very significant. Eventually we identified a dozen experts who had used these technologies to significantly enhance training for high performance individuals, like athletes, firefighters, police units and special force teams, and invited them to the workshop. To these participants we added individuals from the Air Force Training and Education Command, the prime contractors who developed advanced fighter aircraft and of course, current and former F-22 pilots. The workshop was a three-day event with the first day devoted to truly understanding the problem, the second-day developing solution opportunities and the third-day arriving at one or two recommendations that could be implemented very quickly. All three days went as planned and the exercises that we had incorporated into the workshop allowed all the participants to feel that they were heard and valued for their expertise and experiences. On the third day, a program plan was developed to stand-up an experimental training operation to train new pilots using AI, VR and AR technologies incorporated into fighter aircraft training simulators. The plan was briefed to the AFTEC Commander the next week, he approved it and the training program began in less than two months. The first class of pilots trained in these simulators have gone on to actual flight training and the enhancements provided by the advanced simulators are now being measured. Interestingly, one of the data sets analyzed by the students in the Summer of TDA at WBI Downtown was from this experiment. It's clear that this approach will reduce training time, probably not from 6 years to 1 year, but certainly significantly.

Capturing Opportunities is a powerful way to investigate and understand the potential of emerging technologies to a capability area. The innovation and collaboration created in a Capturing Opportunities workshop is guaranteed to produce new ideas and possibilities and that makes the process worthwhile in itself. WBI has engaged in other Capturing Opportunities projects and has seen the same results in very different challenge areas. This approach is an efficient and effective process for any organization to explore possibilities and opportunities of new and emerging technologies on their products or systems.

▌BREAKTHROUGH INNOVATIONS

When the Air Force was created as a separate military service after World War 2, it immediately instituted a science and technology program that would support the development of tactical and strategic technology that would allow the creation of advanced air and space craft systems and guard against technological surprise by an adversary. In the early days of the Air Force, the focus was on the development of superior technological capabilities, particularly to counter the advanced systems being developed by the Soviet Union. In a final push to overwhelm the Soviet Union, President Reagan created the Strategic Defense Initiative (SDI) and the Air Force was a major player in that successful effort. The Air Force led many of the key SDI programs like the National Aerospace Plane and the Brilliant Pebbles Satellite Weapons programs and developed highly integrated prototypes of these weapon systems to prove to the Soviets that the US had that capability. But after the collapse of the Soviet Union, the Air Force research and development program vectored to a broad technology research program with very limited development of advanced systems. AFRL led that effort from 1997 to the present and the costs to cover the research and exploratory development of many potential military technologies that might prove useful against a myriad of potential enemies forced AFRL to severely limit any advanced systems development. In 2018, the Air Force leadership recognize this deficiency and instituted the Vanguard program. WBI has been involved in the Vanguard program since the beginning.

The Vanguard program was the brainchild of the Deputy Secretary of the Air Force for Acquisition, Dr. Will Roper. He felt that the Air Force needed to do something bold to take advantage of the latest technologies and demonstrate the capabilities that could result from those breakthroughs at a systems level. The first Vanguard involved several AFRL engineers from the Aerospace Systems Directorate and focused on autonomous Unmanned Air Vehicles. The objective was to use AI in UAVs for military combat capability. While UAVs have been playing a prominent combat role in the Middle East operations of the Air Force, using autonomous or semi-autonomous UAVs had not been tried. Since current UAVs are controlled by pilots on the ground, they are not autonomous but do have the ability to learn from their history and circumstances. When UAVs with Machine Learning and Artificial Intelligence technologies become even semi-autonomous, the management and control of these systems becomes more complex. Nevertheless, the potential of this capability is so significant that a multi-hundred million dollar program to design, develop, build, test and demonstrate "AI for Autonomous Military Combat Operations" was launched. Like all Vanguard programs, this one was designed to deliver a game-changing capability through testing and demonstration of one or more operational-level systems in three years. The premise was that this would suffice to convince the operational Air Force that this capability was real and that they would then begin to acquire these systems.

In 2018, WBI worked with the AI for Autonomous Combat Operations team to develop a workshop that would bring some of the leading experts and companies in the AI and the UAV world together to discuss what might be possible under the ground rules of the Vanguard program. These rules are particularly challenging because they require actual demonstration and testing of a prototype in three years. In the past, funding for such an advanced development program was always the issue, but with Vanguard, it was no longer the limiting factor.

The workshop was held at Edwards Air Force Base in Palmdale, California, home of the Air Force Flight Test Center. The choice was dictated by the need to get the Test Center on board from the beginning since a three-year program would require actual flight testing in two years and that testing could have a significant impact on the design and operational capability of the autonomous UAVs. By the time of the workshop, the team had begun to use "Skyborg" for the name of the program since it conveys many of the elements of a high-tech air combat capability. The workshop included former Air Force UAV pilots, engineers from organizations that had implemented advanced technologies into systems to provide some degree of autonomy, leading AI development and application experts, test pilots and many engineers and programmers from the Flight Test Center. WBI facilitated the workshop, which included tours of the parts of the test center that were being used for UAV flight testing.

The workshop was successful in defining the technologies that could be developed and incorporated into a prototype UAV in less than a year. Some were discarded because of their low technology readiness level and others because of the challenges in flight testing systems which incorporated these technologies. The workshop concluded with a preliminary agreement on the fundamental elements that would be part of Skyborg and planning for a myriad of future meetings, reviews and workshops needed to make the program successful. WBI facilitators used many of the innovation and collaboration processes that we had developed over the years to help the team with this challenging effort and were involved after the workshop to help with future aspects of the program. It was obvious to us before, during and after the workshop that Skyborg is a great example of Collaborative Innovation. The incorporation and integration of advanced technologies into an unmanned flight system, the pace and high-level visibility of the program, the number of Air Force organizations that will be involved in the program and the pioneering objective of Skyborg necessitates extreme collaboration and innovation. At the workshop, we saw that collaboration take place and simultaneously the creation of needed innovation from that collaboration. As with all Collaborative Innovation, that cycle repeated many times in the three-day workshop to produce what was needed, get agreement, move on to the next challenge, collaborate to produce innovation and so on. Skyborg is a great example of Collaborative Innovation and is a benchmark for all of the current and future Vanguard programs.

Simultaneous with supporting the Skyborg program, WBI was heavily involved in developing the processes, infrastructure and strategy needed to support all of the Vanguard programs. WBI worked with the Air Force Weapons Integration Center (AFWIC) and AFRL throughout 2018-2020 to design and implement the organizational structure, funding processes, technical support and execution elements for Vanguard programs. While the Vanguard approach is still evolving, Vanguard programs like Skyborg are underway and will lead to capabilities that may significantly change the way the Air Force operates. Their success depends on many factors, but Collaborative Innovation will be needed more than ever.

| COLLABORATIVE INNOVATION DEVELOPMENT

As we look back on the many approaches, processes, techniques and tools developed and delivered by WBI to our Air Force and commercial customers, it's gratifying to know that they have been useful in enhancing collaboration and innovation in these organizations. Further, they are intertwined and producing one often generates the other. But we believe there is even more power in the combination of the two, that is Collaborative Innovation. Like any synergistic process, there is more to the combination than the sum of the two. Because of this, we are continuing to research Collaborative Innovation by studying where it has been successful in producing breakthroughs in any organization or circumstance. And we have discovered some amazing things as individuals relate their stories of Collaborative Innovation.

While we believe that Collaborative Innovation is a process that can be applied to any situation, we have found that it really requires a serious commitment on the part of a person or a team to be successful with this approach. As stated in the beginning of this book, it takes energy, trust, steadfastness and belief to truly employ this approach. As we listened to other individuals and organizations that had employed Collaborative Innovation successfully, one commonality stood out. The process was used because there was a break-down or serious challenge of some kind that had to be resolved if an important goal was to be achieved or a disaster averted. That drove the commitment to using Collaborative Innovation. If that commitment was strong enough to see it through, a breakthrough was achieved.

There are many breakdowns and problems in the world of business, government and academia but most don't get resolved into breakthroughs. However, those organizations that do achieve the breakthroughs use a process like Collaborative Innovation to achieve it. So why doesn't everyone use it. There seems to be a few principles and tenets that are at the heart of Collaborative Innovation and we can confirm many of these from observations in our workshops and projects. Six stand out and seem to be prerequisites for Collaborative Innovation:

A constant awareness of the energy in the team and a commitment to create the required energy at the right time,

Human mutuality and respect by all members for each other,

A culture of curiosity and the willingness to actively seek new ideas, information and perspectives,

Absolute honesty in all conversations and commitments, that is, always telling the truth, the whole truth and nothing but the truth.

Actively, openly and caringly listening to each other.

Being willing to be coached and helped if needed in order to reach the objectives, even if it means changing your mind,

When we introduce these principles to teams that have asked us to help with Collaborative Innovation, we often get a surprised reaction. Most people truly believe that they abide by these principles, at least most of the time and under most circumstances. For Collaborative Innovation to really produce a breakthrough, it's about all of the time and in all circumstances. That's very hard because people will rationalize and drift away from these commitments as the going gets rough. It's not unnatural or even bad, it's just that it is tough to do it all of the time. So that is why we are absolutely convinced that you need a trained facilitator to help in this endeavor. It's possible that someone in the group can play that role, but in doing so they often have to give up the role or responsibility that they would normally play. For a leader, this means relinquishing the role of being the leader and that is often not possible or even appropriate. At WBI, we usually insist that one of our certified or trained facilitators is used if Collaborative Innovation has been requested by the customer. Like very tough legal, business or domestic situations use counselors or consultants to get to breakthroughs, Collaborative Innovation needs similar support to help the players and participants get to an amazing outcome.

To understand the power and possibilities of Collaborative Innovation and to increase the number of trained facilitators in this practice, WBI has partnered with Collaborative Innovation Institute to conduct significant research in this area, train and certify CI facilitators and offer Collaborative Innovation programs to for-profit and non-profit organizations locally, nationally and globally.

COLLABORATIVE INNOVATION APPLICATIONS

"An idea that is put into action is much more important than the idea itself", Buddha

▎ THE ART AND SCIENCE OF COLLABORATION INNOVATION

It would be convenient if Collaborative Innovation were simply a process that could be applied to a challenge or problem in order to generate solutions or possibilities. Unfortunately, it's not that simple. As we have seen at WBI, Collaborative Innovation depends on a strong collaboration between the individuals involved and there is an art to achieving that. Like any significant collaborative endeavor, human interaction and synergism are involved in Collaborative Innovation and managing that is both an art and a science.

A symphony orchestra is an assemblage of dozens of professional musicians playing the notes of the selected musical piece in harmony. Without a conductor, the orchestra will produce good music, but the work of the conductor, both before and during the performance, is to stimulate and motivate the orchestra to play not just good but great music. Great conductors have mastered the art of symphonic collaboration and integrated that with the process of producing the sounds that are required to deliver the piece. Similarly, sport coaches are expected to do the same, that is, foster intense collaboration between the players as they execute the plays in the game. In both cases, a great amount of effort is required before as well as when the orchestra or team performs. While knowing the principles and practices of collaboration will help the conductor or coach, the mastery of collaboration is an art which comes from expertise and experience. Similarly, Collaborative Innovation requires the same mastery if it is to deliver great results.

Usually, leaders are taught to encourage and motivate their teams to be collaborative. Much time and effort is spent in most organizations to achieve collaboration. Unfortunately, it is often seen as a separate activity or element from the business of the organization, which is usually governed by processes and practices. This separation minimizes the power of Collaborative Innovation because collaboration and innovation are not treated holistically. Even the most connected team can stray from a collaborative effort when people disagree or the process requires some give and take. That breakdown, while sometimes subtle, can be extremely detrimental to achieving great results. If Collaborative Innovation is to be achieved, that breakdown must be managed immediately and aggressively. And that requires help from someone who is trained and focused on doing just that. While the leader might be able to do that, often that person can't and almost always, that person shouldn't. A special talent is required and that's where the art comes in. In sports, that person is the coach. In business, we call them facilitators.

FACILITATING COLLABORATIVE INNOVATION

We have found that the facilitation of Collaborative Innovation is probably the most essential element of the process. With all good intentions, teams congregate every day to develop solutions to critical challenges or change the way that the organization is doing business. The leader usually sets the objectives for the meeting and the participants are urged to work together to develop potential ideas or solutions to deal with the challenges. Even with the most collaborative teams, it doesn't take long before differences of opinions or perspectives rear their heads and the meeting turns into a competition of ideas, often personalities. The best leaders keep pushing back on this competitive element, urging the participants to collaborate. But the leader is in a very awkward position to do this. Pushing back on any idea or interchange looks like the leader is against this, encouraging any idea or interchange looks like the opposite. Yet the leader can't be neutral and remain a leader, nor can the leader let the discussion degrade to a competition or standoff. The only solution is to get help from a third party, a facilitator whose main job is to encourage and maximize collaboration while the team tries to generate the ideas and possibilities that are required for the challenge.

Leaders are often reluctant to do this. Whether it's ego, loss of control, lack of trust or simply the resources required to train and utilize a facilitator skilled in the art of collaboration, leaders generally don't utilize facilitators in most of their meetings. So it is important for leaders to understand that facilitators can be partners to them, just as their administrative assistants, seconds-in-commands, trusted aides and number two's support and assist them in their mission. Most leaders would never hesitate to bring a lawyer to a negotiation meeting, a financial expert to an audit or a personnel specialist to a human resources meeting. Yet they enter into a meeting with their peers and subordinates with no specialized helper in this area. Facilitators are the key to successful group meetings and Collaborative Innovation facilitators are critical to the success of Collaborative Innovation endeavors.

While there are a variety of programs and trainings that focus on developing the facilitation skills of individuals, Collaborative Innovations facilitators are a special set of this practice. While most facilitators focus on helping with the mechanics, agenda and actual content of a meeting or workshop, Collaborative Innovation facilitation requires an additional talent; they must manage the context of the meeting. To do that, they must understand the intended context of the meeting and do everything they can to get everyone to support that. For all Collaborative Innovation efforts, the context is collaboration and not innovation. If the group can achieve and stay in collaboration, innovation will follow. The primary job of a Collaborative Innovation facilitator is to make this happen. Our research and experience has shown that there are six principles that assist in this endeavor and we have trained our Collaborative Innovation facilitators to be particularly sensitive to these principles.

The first principle is energy awareness, being intentionally attuned to the energy of the group and to its effect on collaboration. While high energy may be appropriate for some aspects of the effort, there are times when reflection, introspection and individual self-awareness are required. The facilitator must be aware and help manage the energy of the group. This requires almost-constant attentiveness to the individual and collective energy in the activity and the deliberate, and sometimes courageous, actions that must be taken to keep collaboration on track.

The second principle is human mutuality. In general, the individuals in a meeting or workshop will behave well, show respect for each other and may even know, like and count on each other. But human mutuality is beyond just getting along, working with each other and civil behavior. Human mutuality is the acknowledgement that we are all in this together and that the outcome depends on the collaboration, synergism and mutual dependence of everyone on the team to achieve the agreed-to goal. The facilitator must periodically, sometimes constantly, remind the team of this and make this a part of the context of the meeting. Again, this requires attentiveness to the team behaving as a team as well as dealing with disruptions to team behavior by individuals who are more interested in their own agendas than that of the team.

The third principle is curiosity. Many individuals are curious and that very attitude helps the team search for new approaches and solutions. But collaboration is difficult and collaborative innovation is even more challenging. It's easy for a team to get bogged down and then begin to spiral towards complacency or even dysfunctionality. The facilitator must anticipate this and watch for any sign that the team believes that there are no new ideas or approaches or prospects for achieving their goals. When that happens, the simplest technique is to begin asking questions in order to open up the conversation to new ideas and suggestions. There are times in a meeting where convergence is absolutely appropriate. But if convergence is too premature or simply a reaction to time pressures, exhaustion or a reasonably good solution, it must be challenged. Divergence might be difficult to introduce at times when the group seems intent on moving on, but it likely will yield a better solution or even a completely new approach to the challenge. The facilitator's job is to catalyze and stimulate curiosity in the group and to do whatever it takes to achieve it.

The fourth principle is committed speaking. In our experience, committed speaking is the most important principle necessary to achieve Collaborative Innovation. Most people speak the truth and believe that they are intentional about what they say. However, it's easy for all of us to consider a conversation as simply a communication of words and sentences and not as a solid commitment to follow through on what we are saying. This is particularly true for those "extroverts" who talk before thinking, less so for the "introverts" who think through a thought before speaking. Committed speaking certainly allows for the interchange of ideas, possibilities and opportunities in a casual way. However, once a path, plan or follow-on activity is sincerely proposed, the reaction of every member of the group to that idea requires committed speaking if collaboration is to follow. When things get serious, the individuals in the group must speak from the heart and be committed to their contribution to the dialog. The Collaborative Innovation facilitator's role is not to question anyone's contribution to the discussion but to watch to see if people are truly speaking with commitment. At first, everyone is likely to believe they are committed to what they say. But by closely examining the dialog, the facilitator must be able to determine what really is going on and must challenge those individuals who do not seem to be committed to what they are saying. This is a difficult job and a hard call to make when something seems amiss. But it is crucial for the facilitator to be attentive to this. Otherwise, the meeting ends with a belief that the objective has been met while some participants are only tentatively committed to the results. By having committed speaking as a tenet of the conversation, much deeper discussions and much stronger commitments will result. True collaboration occurs when the collaborators are committed to what they say and are willing to stand by their words. That type of collaboration then opens up the possibility of surprising dialogs, deep discussions and true innovation.

The fifth principle is generous listening. We have all been taught to be open-minded and active about listening to others. Generous listening goes beyond that and is a way of creating trust in the group which then supports curiosity, intention and humanity. It is beyond hearing what someone is saying. It is a way of being when interacting with others. Generous listening starts from a place of fully valuing people and a commitment to understanding what is important to them. By practicing generous listening at all times and insisting that everyone else in the group follows suit, Collaborative Innovation facilitators can transform meetings and workshops and the team can achieve breakthrough innovations. At the appropriate time, facilitators should surface thoughts and feelings that are getting in the way of generous listening and encourage the group or individuals in the group to let that go. Collaborative Innovation facilitators must listen and engage with the group from that context and perspective and do whatever is needed to transform the meeting to a collaborative innovation opportunity.

The sixth principle is coach-ability. This principle is really a two way street. The ability to be coached certainly applies to the facilitator and a Collaborative Innovation facilitator must seek and create opportunities to be coached. Even in the midst of a workshop or meeting, it is important for the facilitator to solicit and be welcome to opportunities to be coached to do a better job. Sometimes this comes in the form of suggestions from a participant or leader but more often from a keen awareness of what is really going on and reacting appropriately to that. Primarily however, the ability of the group or any of the participants to be coached, particularly by the facilitator, can make or break a meeting. If the group is not open to this, then the facilitator's job to achieve collaborative innovation is an uphill battle. At the outset, the facilitator should position his/herself as a coach to the group. The analogy of a sports coach can be used to convince the group to see the facilitator as a coach. In sports, as well as many other individual or group programs, the coach helps the group achieve maximum performance. Nobody would argue with the need for a coach in sports, theatre or education, but a business team generally doesn't see the need for a coach to enhance their performance. The facilitator is ideal to play this role and acknowledgement of the facilitator and the participant's roles in this is a great way to start the meeting or workshop. Certainly there may be other coaches needed during the activity, for example, a person to coach the team towards innovation. But again, powerful collaboration is difficult to attain and maintain and it is the primary role of the Collaborative Innovation facilitator.

By utilizing these six principles, a Collaborative Innovation facilitator can help a team and workshop achieve what is needed to turn breakdowns into breakthroughs.

❙ INVENTING POSSIBILITIES

One of the most rewarding applications of Collaborative Innovation is Inventing Possibilities. We have conducted a number of Inventing Possibilities workshops at WBI and they are very exciting. They start with a focus on a broad opportunity or capability area, such as, training, human performance or covert operations. The central theme could also be even broader, like climate control or bio-terrorism. We then carefully develop a potential group of participants from both the problem space and the potential solution spaces. For the former, individuals who have been working in the focus area or who have a great amount of experience in that field are identified and invited to the workshop. The potential solution space participants are more challenging since a very large number of possible solution areas might be applicable to the focus area challenge. The most general Inventing Possibilities workshop that we have conducted started with identifying the fifty most exciting new technologies or capabilities of the next decade. There are a number of global organizations who carefully assemble such lists and they can be used as a starting point. While it would be impossible to assemble fifty global experts for a workshop, we have used techniques to narrow down the list, such as clustering, pairing technologies to the focus area and even random selection from the lists. Our experience has shown that you don't need (and can't accommodate) everyone that might be of interest to have a successful event. Having several participants to represent the focus challenge and 10-20 experts representing technologies of the future is usually sufficient to have a great workshop.

The secret is to spend a great amount of effort in preparing for the workshop and using Collaborative Innovation facilitation and processes during the workshop. While we refer to these events as workshops, they may require several workshop of multiple days to accomplish. A significant amount of time, perhaps even several preparatory meetings or workshops, is spent on the collaborative element of the process. Ideally, the participants are fully engaged, committed, trusting and bonded with each other before any attempt to innovate in the focus area begins. Amazingly, it doesn't take much time or effort for the participants to reach a very high level of intensity and energy. Most people are not only willing to entertain possibilities but often really enjoy discussing, creating and collaborating with each other to generate possibilities. Often the energy is so high and the conversations so engaging and exciting that the facilitator's main job is to harness the energy in order to capture the ideas and possibilities.

The Inventing Possibilities workshops almost always generate too many possibilities or opportunities to even consider implementing. This is very similar to the divergent phase of a creative problem solving activity and various methods of converging, prioritizing or narrowing the results must be used. We have used many approaches to converge and the participants are usually very receptive and engaged in that part of the workshop. The objective is to create a few breakthrough possibilities for further consideration and these generally come in the form of unique capabilities that integrate very advanced multi-disciplinary technologies or systems. For example, combining advanced multispectral sensors with flexible nano-material clothing, machine learning Artificial Intelligence, 5G information management and micro quantum computers could produce uniforms that essentially make warfighters invisible, certainly an interesting possibility.

While Inventing Possibilities workshops can be used to generate capabilities and opportunities that could greatly change the future, they can also be used to examine and generate possible futures. Assembling experts from a wide variety of fields and professions, such as finance, medicine, governance, business, education, information technology, transportation, etc. and using Collaborative Innovation facilitation and techniques to examine future scenarios, such as urban development, international collaboration, space law, trade agreements, etc. But the key to Inventing Possibilities is to facilitate the activity such that collaboration is maximized and managed in order to stimulate and generate breakthrough ideas and possibilities.

▌HARNESSING OPPORTUNITIES

When you consider all the advances being made in science, technology, medicine, human performance, business, finance and so many other fields, it's a bit overwhelming to see how these might apply to the future of your organization and people, your products and services or your capabilities and operations. Harnessing Opportunities is a process that we have used to narrow down the possibility and opportunity space to a more manageable situation. As we learned in our workshops on Capturing Opportunities, starting with a "hunch" or possible vision and then examining potential new innovations that might really change the way that vision could play out can yield amazing opportunities. There are untold opportunities out there but which ones might be useful to an important challenge or possibility. Start with a hunch, a gut feeling or a brainstorm that you feel has merit. The starting point will likely point to some innovations or trends that might allow the capability to grow in a powerful way. The process is then to look at these innovations with respect to how they might impact the capability if they are applied in creative ways.

Several years ago, we examined the possibility of applying Artificial Intelligence, Augmented Reality and Virtual Reality to high-performance aircraft pilot training. By carefully researching these areas, we found that there were companies and organizations that had applied these three technologies to enhance training for a variety of operations, such as first responders, fire fighters, truckers, FBI agents and private airplane pilots. We invited representatives from these companies to a multi-day workshop which also included high-performance aircraft pilot training experts as well as scientists and engineers who were carrying out research and development on state of the art Artificial Intelligence, Augmented Reality and Virtual Reality. By following the practices of Collaborative Innovation, we developed a number of approaches to pilot training that significantly reduced the time necessary to train a pilot and also enhanced the performance potential of that pilot. Within six months after that workshop, several prototype training programs were developed and then tested in actual Air Force pilot training programs. The results were outstanding and more work is underway to enhance high-performance aircraft pilot training using these three advanced technologies.

By examining the potential of advanced technologies, breakthrough innovations and novel trends on current capabilities, new opportunities can be captured and then harnessed to develop breakthrough possibilities and powerful capabilities. The key is to manage the process to achieve the maximum collaboration, first from the technology, innovation and trend participants, then with the entire group. By focusing on the potential multi-disciplinary possibilities and presenting these to the capability owners, new possibilities and opportunities will present themselves. This interaction can be iterated many times until new possibilities are generated. Throughout these cycles, incorporating comments from current practitioners can be solicited. The current practitioners may be reticent or even against new possibilities for a number of reasons but that shouldn't negate their involvement. Again, by constantly assessing and energizing the collaborative nature and commitment of the group to true collaboration, surprising innovations will result. Harnessing opportunities like this will not only provide an organization with new and improved products and services but will also develop very strong collaborations within the organization which might yield further breakthrough innovations.

▌ FLIPPING BREAKDOWNS

A breakdown is usually not a desired goal, but it might provide just the right opportunity to create a breakthrough. When things go really bad, a project is in serious danger of being cancelled or a team becomes seriously dysfunctional, the normal reaction is to give up and move on to something else. But it is the perfect time to see if you can flip it to a breakthrough. Two of the key principles of Collaborative Innovation are energy awareness and speaking the truth. If the project or team has hit its lowest point, the energy of the group will be low. On the other hand, what have you got to lose by examining how you arrived at this situation? The breakdown provides a great opportunity to look at what went wrong, how the team moved to this point and what caused the breakdown. If the team just moves on, there are no lessons to be learned and the same situation might occur again. Conversely, by examining the energy flow of the group that ended in the breakdown and trying to really understand with open, caring dialog and generous listening what led to the breakdown, the key decisions and behaviors that need to be flipped can be identified. During some of our most successful Collaborative Innovation projects, breakdowns have occurred. But we didn't give up, we tried to understand what had caused that result. By examining the behavior of the group with respect to the six principles, we began to understand what had happened. Was there a point or period when the energy of the group was seriously compromised by other demands or were there some individual behaviors that sucked the energy out of the team? Did mutual respect and support change dramatically? Did the team get bogged down in the details such that curiosity was lost? Was truth-telling put aside or was the team reluctant to deal with the truth? Did some or most of the team stop listening to each other or become so vocal with their ideas as to silence others. And were constructive criticisms and new ideas seen as attacks? These are all breakdowns in collaboration and need to be reversed or corrected if the team is going to be successful. Usually a trained facilitator can pick these up before the breakdown occurs and take measures to reverse it before the team collaboration collapses. But if it does happen, it's a perfect time to identify the issues, have the team acknowledge them and work on getting back to collaboration. While it's not easy, it can completely reverse the direction of the team because now they know what to look for that is getting in their way.

Fortunately, we've had quite a few projects and workshops that have flipped breakdowns into breakthroughs. Often a group or team asked for our help when they had a breakdown and wanted or needed to reverse this situation. We start by looking at any deviation from the six principles of Collaborative Innovation and try to get the team back on track. Sometimes they are not even aligned in what they really what to achieve and we apply organizational alignment facilitation as well. But when we can flip a team from a breakdown, the results are usually quite amazing. Just like in our personal lives, coming out of a low point with support and assistance can lead to breakthroughs. The same can happen with teams and projects and using the principles of Collaborative Innovation can be extremely useful to accomplish that change.

▌ TRANSFORMING ORGANIZATIONS

There's a famous organizational development tenet that says, "Culture Trumps Everything". It's true, but that doesn't mean that change can't happen. Maybe the culture needs changing.

Since Wright Brothers Institute is an Innovation and Collaboration Institute, literally hundreds of teams and organizations have come to us for help because their teams or organization were not innovative or collaborative. When they do, we usually examine the processes, structures and strategies of the organization to see if there is good alignment or if there are barriers or bottlenecks getting in the way. We call this process Organizational Alignment. If we can see where there is non-alignment, such as divergent strategies, dysfunctional structures, conflicting processes or meaningless reward systems, we can quickly focus on the problem and recommend some fixes or changes that will help. But if culture is the culprit that is getting in the way of collaboration or innovation, that's more challenging. Unfortunately, often it is and some change in the culture will be necessary before desired results can be achieved.

Culture change is very difficult and often takes great effort and a long time to achieve. Just wishing or demanding that the organization's culture can become more collaborative or innovative doesn't work. And yet, leaders don't want to spend the time or energy to implement an organizational culture change program. They want it to happen fast, sometimes right away. While we never promise that we can achieve a fast and easy culture change, we have used Collaborative Innovation to achieve some rapid results and "quick wins" in innovation and collaboration. Again, we employ the six principles with a good measure of tough love and hard questions. Can you feel positive energy or is the group resigned to defeat? Is there mutual respect for everyone in the organization or is it an "us/them" culture? Are people really curious or do they think they know it all and aren't open to new ideas? Do people really tell the truth to each other or are they afraid that they would suffer if they did? Do members really listen to each other or just pretend to do so? If we can get the team members to truthfully answer these questions, change can happen fast. It takes courage for everyone to be involved in such a discussion, particularly the leader. But it can be an incredible effective way to change the direction of a team and even a large organization towards collaboration and innovation. We certainly agree that a team should not "try this at home", particularly without a trained facilitator who understands Collaborative Innovation principles. But it can and has worked with dozens of teams and organizations that we have helped. While an instant change to living by the six principles never happens, it is a start and with continued attention to moving in that direction, collaboration will be improved and innovation will follow. Trying to improve innovation in an organization by campaigns, innovation training, innovation processes and innovation awards usually doesn't get the job done. We found that innovation requires collaboration and working on getting the organization to be more collaborative through the six principles gives you better and quicker results.

| CULTURAL COLLABORATION

Because the primary mission of Wright Brothers Institute is to be a "partnership intermediary" between very large government, academic and business organizations, we try to stimulate, catalyze and facilitate collaboration and innovation between particular elements of these organizations. That's even trickier than getting a single team or organization to move their culture towards innovation and collaboration. From the very first project that we took on, trying to form a collaboration between the country's three major aerospace prime contractors and the government, we faced this challenge. Each of the three companies had different cultures. For example, Boeing has a major commercial aircraft business as well as a military aircraft systems business while Lockheed

Martin and Northrop Grumman have broader military aerospace systems capabilities. While they are all huge companies with some common and competitive goals, their cultures are not the same. When the leaders of these companies met at WBI to discuss forming a collaboration on pre-competitive simulation testing, it was clear that we were talking to three different cultures. When we later integrated the Air Force mission simulation community into conversation, another very different culture was at the table. At the time, we hadn't researched and developed all the principles of Collaborative Innovation but we did know some of them. At several meetings, a talking stick was passed around and the rules of the talking stick, generous open listening and speaking from the heart, were observed. This led to some innovative possibilities for collaboration and we used these to continue the discussion for collaboration opportunities. There were other instances when the energy of the group for those possibilities waned and we actually paused the meetings to engage in more social activities in order to regain the momentum. We did brainstorming around possible opportunities and the curiosity of the group was piqued. As facilitator, I kept reminding the group that we were all on the same side and committed to doing the best for the country and that helped renew the mutuality of the various players. So even without knowing what we know now, we used the Collaborative Innovation principles and formed a successful collaboration around pre-competitive testing and simulation that saved all of the companies and the government millions if not billions of dollars.

That old example just reinforces the power of Collaborative Innovation when different cultures have to collaborate and innovate. The same six principles can be used to open the gates of different cultures to allow meaningful discussions and innovative possibilities. Like all of our previously cited examples, once possibilities emerge, opportunities will present themselves. If real collaboration can be maintained, those possibilities will turn into opportunities and great agreements can occur between very different cultures.

| CHANGING THE WORLD

While not in the specific mission of the Wright Brothers Institute, everyone at WBI would love to contribute to changing the world for the better. There are so many challenges that are outside of our scope and focus, but there is no reason why the principles and practices that have worked for us couldn't be applied to the world's biggest problems. Most of these challenges involve dilemmas and conflicts between groups with very different perspectives. Climate change pits environmentalists against big businesses. World hunger hits at the core of who is responsible for the survival and development of much of the world's population. Many military conflicts can be traced to widely different interpretations of the same beliefs by different religious groups. These are clashes between different cultures and points of view and these differences are very difficult to resolve. Nevertheless, every attempt needs to be made to achieve some progress and the only hope is to get the parties to collaborate and innovate. Using Collaborative Innovation, new possibilities and opportunities can be generated and hopefully some of these can be acted upon to change the world for the better.

Printed in the United States
By Bookmasters